# ENGINEERING IN PERSPECTIVE

## Lessons for a Successful Career

## Tony Ridley

Imperial College London, UK

 **World Scientific**

NEW JERSEY · LONDON · SINGAPORE · BEIJING · SHANGHAI · HONG KONG · TAIPEI · CHENNAI · TOKYO

*Published by*

World Scientific Publishing Europe Ltd.

57 Shelton Street, Covent Garden, London WC2H 9HE

*Head office:* 5 Toh Tuck Link, Singapore 596224

*USA office:* 27 Warren Street, Suite 401-402, Hackensack, NJ 07601

**Library of Congress Cataloging-in-Publication Data**
Names: Ridley, Tony, 1933–    author.
Title: Engineering in perspective : lessons for a successful career /
    by Tony Ridley, Imperial College London, UK.
Description: New Jersey : World Scientific, [2017] | Includes bibliographical references.
Identifiers: LCCN 2016037119| ISBN 9781786342270 (hc : alk. paper) |
    ISBN 9781786342287 (pbk : alk. paper)
Subjects: LCSH: Engineering--Vocational guidance. | Project management--Case studies. |
    Construction projects--Case studies. | Transportation engineering--Case studies.
Classification: LCC TA157 .R56 2017 | DDC 620.0023--dc23
LC record available at https://lccn.loc.gov/2016037119

**British Library Cataloguing-in-Publication Data**
A catalogue record for this book is available from the British Library.

Desk Editors: Anthony Alexander/Mary Simpson

Typeset by Stallion Press
Email: enquiries@stallionpress.com

**Professor Tony Ridley, President of ICE, 1995–1996**

By kind permission of Michael Noakes, Portrait Painter, and ICE

# About the Author

Tony Ridley is the Emeritus Professor of Transport Engineering in the Department of Civil and Environmental Engineering at Imperial College. He was Head of the Department of Civil and Environmental Engineering (1997–1999) and a Professor within the department since 1991. He founded and was the Chairman of the Railway Technology Strategy Centre through which he was involved with two Metro benchmarking clubs, CoMET and Nova, which include more than 30 of the world's leading Metro systems. He has acted as Consultant/Advisor to Metro developments in a number of cities, including Bangkok, Jakarta and Singapore.

He joined the Board of Eurotunnel as a non-Executive Member in 1987. He was Managing Director of the project for two years. Immediately prior to that he was eight years on the Board of London Transport and was Chairman and Chief Executive of London Underground Ltd. He was involved with the Docklands Light Railway from the outset, latterly as Chairman.

He was the first MD of the Hong Kong Mass Transit Railway Corporation (1975–1980) and the first DG of the Tyne and Wear Passenger Transport Executive (1969–1975).

He holds a Doctorate in Transportation Engineering from the University of California as well as Civil Engineering degrees from the University of Newcastle and Northwestern University. He attended the Senior Executive Program at Stanford Business School in 1980.

He is a Fellow of the Royal Academy of Engineering and of both the Institution of Civil Engineers (of which he was President 1995–1996) and the Chartered Institute of Logistics and Transport (of which he was

International President 1999–2001). He became President of the Common wealth Engineers Council in March 2000 and was a member of the Executive Council of the World Federation of Engineering Organizations. He is a former President of the Light Rail Transit Association and of the Association for Project Management. He was a Director of the Major Projects Association. He has been active in the International Union of Public Transport and is now an Honorary Member.

In 1988, he was the first recipient of the Highways Award of the Institution of Highways and Transportation and, in 2000, the first recipient of the Herbert Crow Award of the Worshipful Company of Carmen. In March 2002, he received the President's Award of the Engineering Council. He has been a Trustee, and a member of the Public Policy Committee, of the RAC Foundation for Motoring.

He was a member of Task Force 10 (Science, Technology and Innovation) of the UN Millennium Project. He was Senior Transport Advisor to the London 2012 Olympic Bid, and a member of the Olympic Development Authority's Independent Dispute Avoidance Panel. He was Chairman of the Investments LLP for the Government's (£multi-billion) Building Schools for the Future programme (2007–2010).

# Acknowledgements

*Engineering in Perspective* is derived from more than 60 years of endeavour on my part, but it has greatly benefitted from lessons I have learned from educators and students I have taught, from colleagues I have worked for and those who have worked for me, and from other professionals and politicians.

The book is dedicated to the memory of Professor Fisher Cassie, who first introduced me to the breadth of engineering, but I am also indebted to many important actors in my life who are recognised by name, as are personal mentors.

This may be my last 'major project', and I acknowledge a small number of people who have made particular contributions: Roger Allport, with whom I worked in a number of cities in Southeast Asia and who has read countless versions of chapters, Mike Chrimes and John Gold, as well as others who have offered insights on individual chapters. It was always my ambition to find a truly helpful editor whom I could thank sincerely and in Jane Sayers at WSPC I have done so, with other of her colleagues at WSPC.

My family has been very supportive, not least Jane Ridley who is still one book ahead of me. It is fashionable to talk about the 'love of my life'. She really has been, but also my 'carer' during hard times and a *critic extraordinaire* when necessary.

Tony Ridley

# Contents

# 1

# Introduction

## 1.1 Aims and Method

This book aims to identify a number of propositions, developed during my professional career, that relate to the nature of engineering and, consequently, the education and training of engineers. Engineers do much good for society but we have focussed too much, in the past, on our products, rather than the services that our products provide for society.

Civil Engineering projects have three essential stages:

- What shall be done and why, and where is the money to come from?
- Design and construction.
- Service operation which meets society's requirements, and the objectives of the project.

Far too many engineers have regarded only stage two as their concern. But repeatedly through my career I have discovered that 'there's more to engineering than engineering' which draws attention to the breadth of knowledge required.

Today's engineers must recognise that many of the skills needed are not based on the traditional maths and physics, important though those are. In the Epilogue, I ask educators and employers to meet the challenge of developing young civil engineers who will provide the totality of services that the public require through the provision of infrastructure, which I paraphrase as 'things'. Other branches of engineering have their own 'things'.

My professional career was blessed, in that the first two managerial responsibilities I had were for two projects (Tyne and Wear Metro and

Hong Kong MTR) in very different circumstances, but both were founded on clarity of objectives and sound project planning. I also had a very wide series of other engineering experiences, which developed my thinking — the birth of the Greater London Council and the London Transportation Study; the London Fares Fair row, followed by a 60 percent increase in ridership on the Underground in five years; the King's Cross fire; the creation of the Docklands Light Railway (DLR); the furious battles over the Channel Tunnel project; and winning the London Olympic bid.

Beyond projects that I treat as case studies where experiences taught me so much, I was additionally blessed in having been invited to take a Chair, and then Head of the Civil Engineering Department at Imperial College, which gave me an opportunity for extended reflection. It was apparent that very bright students were eager to learn about examples of engineering drawn from the 'real world', about which I was able to give them first-hand knowledge. They were fascinated to compare the Hong Kong MTR, as a 'case study' of good project management, with Eurotunnel which depended on some superb technical engineering, but had many destructive managerial problems during its construction.

The term 'there's more to engineering than engineering' seeks to examine the wider challenges of engineering projects which demand knowledge and skills that go far beyond the academic and scientific base of their education. For those of my generation, from the mid-20th century onwards, engineering was far too often treated only in its 'techie' context, with the result that engineers both failed to deliver as much as they might have done, and were less influential in society than they might and should have been.

I explore engineering from the perspective of my professional experience, and include some lessons learned, which varied from project to project.

## 1.2 The Breadth of Engineering and Hardy Cross

Professor Fisher Cassie was Head of the Department of Civil Engineering at King's College, University of Durham (in Newcastle) in my time. To me, he was the supreme mentor among people who helped me during my career. When I gave the inaugural William Fisher Cassie lecture (Ridley T, 1995) I closed by saying,

> 'To whom shall we turn for guidance? I suggest that, while we all have our own mentors, there is no better guide than the man who taught us to write,

taught us to speak, was international in his outlook, worked within many disciplines, encouraged new disciplines, thought about the needs of tomorrow, was multi-professional, who had an art in teaching, believed more in mastering the subject than the requirements of exams, who abhorred the division between the academic and the practical, wrote with the needs of students in mind and who, above all, saw and rejoiced in the diversity of Civil Engineering. Such a man was William Fisher Cassie.'

Cassie required his students to write a term paper on our profession. Mine, 'The Aims of an Engineering Education' (Ridley T, 1954), suggested that 'a wider outlook is necessary for the engineer of the future, for his own good, and for the good of society', a belief that has stayed with me ever since. Cassie encouraged such a view, as did Hardy Cross of 'moment distribution' fame (1930) who was well known to students of structural engineering of my generation, in his somewhat philosophic 'Engineers and Ivory Towers' (Cross, 1952).

**PHOTO 1.1.  Professor Fisher Cassie, a great Educator.**

*Source*: (William) Fisher Cassie, By Bassano Ltd., Half-plate film negative, 21 December 1961. Credit: National Portrait Gallery, London.

'It is customary', he wrote, 'to think of engineering as part of a trilogy — pure science, applied science and engineering. It needs emphasis that this trilogy is only part of a triad of trilogies into which engineering fits

- First — pure science, applied science and engineering.
- Second — economic theory, finance and engineering.
- Third — social relations, industrial relations and engineering.

Many engineering problems are as closely allied to social problems as they are to pure science. The workaday world does not fit into an academic department or into so-called fields of learning. It is the whole man who works, the whole community in which he lives. Those who devote their lives to engineering are likely to find themselves in contact with almost every phase of human activity. Not only must they make important decisions about mere mechanical outline of structures and machines, but they are also confronted with the problems of human reactions to environment and are constantly involved in problems of law, economics and sociology'.

PHOTO 1.2.   Prof Hardy Cross, University of Illinois.

This was the engineering that I wanted to engage with. Experience has, time and time again, confirmed the wisdom of Hardy Cross's words.

But Hardy Cross was not alone. My paper drew on the eminent Indian Railway engineer, manager and commissioner, 1897–1928, ICE Past President Sir Clement Hindley's statement (1941) that, 'in carrying through an engineering project, probably 75 percent of the directing engineer's time, energy and brainpower are absorbed in overcoming human difficulties and the remaining 25 percent in solving material and physical problems. Furthermore it is necessary for the engineer to recognise *at the very outset of his career* that, on the one hand, his work will be influenced profoundly by sociological and political considerations and, on the other, that he must make himself acquainted with the fundamental aspect of these considerations if he is not to occupy a permanently subordinate position in the community.'

Professor Sir Charles Inglis's ICE Presidential Address (1941) said that 'too often one encounters the young man who assimilates with meticulous diligence every scrap of knowledge imparted to him and, in consequence, passes all his examinations with inevitable precision. His immediate advancement is thereby assured and it is an absolute certainty that his diligence will carry him up to a certain level, but it is more than likely that lack of humanity and *breadth* of outlook will put a limit on any other advancement.'

## 1.3 Mentors

A variety of people appear in my story, some eminent, others unsung heroes. I believe in the importance of 'mentors' — people who greatly guide and influence us. They probably do not recognise this role. It is only later that the 'mentored' realise who their mentors have been, and perhaps the mentors never do. Each of their names appears at the appropriate chapter.

I learned the importance of *perseverance* from both parents, which came in useful when I was struggling with my PhD dissertation. My headmaster encouraged 'breadth in education'. As a shy teenager, I grew up 'never to be afraid to ask'. By the time I set off towards my eventually chosen specialization, I had experienced pre-stressed concrete, fatigue of steel structures, geotechnics, and hydraulic models, and could claim 'muddy boots' experience too. To these were added debating and acting as life skills. A Native American Indian in the Pacific Northwest was the first person to show me that 'there's more to engineering'.

## 1.4  The Totality of Engineering Projects

It was increasingly clear that an understanding of the *totality* of engineering projects must include the following, which are *absolutely essential* for senior engineering managers, and *highly desirable* for all good engineers. The purpose of engineers is to serve society; the purpose of engineering projects is to deliver the services that society requires, and the success of projects needs to be judged over their operational lifetime, based on how successful they are in delivering sustainable services. The starting point should be what society requires. However, the following should be recognised and understood:

- The supply side of the equation: how to achieve this is *too often the sole focus* of engineering, but is in fact the subsequent issue to understanding purpose.
- The need for a project to deliver expected benefits to customers in operation, as well as completion 'on time and within budget' in construction.
- The profound complexity of major projects, and the need to master all the relevant disciplines, including the politics, economics, finance and communication.
- The crucial importance of clear objectives, the context of a strategic plan, an intelligent client, and sound decision-making processes.
- The critical importance of the interpersonal skills, in managing stakeholder expectations, communicating with the media, leading strong teams and achieving good relations with the supply chain.

    The various chapters that follow illustrate examples of these and other considerations that are prerequisites for engineering success, way beyond the narrow view of what engineers do.

## 1.5  Finding a Way around the Book

The stories I tell are not only for students wondering about their future direction, engineers embarking on their careers, but also for their educators and employers, and for those who lived through the times with me. Some lessons learned (not an exhaustive list) that are referred to in the text are listed at the beginning of each chapter.

To support the authenticity of the content of the book, I call on material relevant to each project from papers I have kept throughout my career. Reference to publications, papers, projects and ideas are included in the chapter bibliographies.

**Chapter 2** describes my path towards an engineering career, when education was part of learning lessons. My propositions arise from my first managerial job onwards, starting with the GLC at the age of 32.

Four years with the **Greater London Council (Chapter 3)** gave the opportunity to put into practice new knowledge of transport analysis and engineering, but taught the danger of focussing on computer-based numbers ahead of 'thinking', and of pressing on with a project without properly addressing affordability. It also began to come clear that *'transport is politics'*.

**Tyne and Wear Metro (Chapter 4)** was blessed with a Minister who, with her civil servants, saw clearly that well-conceived *objectives, policy statements and plans* are, most often, central to success — but with well-planned management of implementation too. Planning of cities and planning of transport go together. But a major project — however well planned and organised — often succeeds only because, fortuitously, the *time is ripe*. The crucial requirement of *political consensus* can often make all the difference between success and failure, but engineers should not leave that consensus solely to politicians.

In **Hong Kong (Chapter 5)**, the MTR was the epitome of the importance of *clear objectives*, based on an understanding that there are no single or simple solutions to transport problems (as also in Singapore). It was a project with *a champion*, an excellent and *intelligent client*, and attention to *good process and decision-making*, together with a project team with a strong team spirit and sense of responsibility, thus able to manage great complexity.

The task with **London Transport (Chapter 6)** was a *rescue mission* after years of decline. The basic challenge was the *management of change*, generally agreed to be yet more difficult than the most complex infrastructure project. It covers the reversal of years of declining ridership to rapid growth and the development of a long-term strategy that paid dividends into the 21st century. Later, **Management after a Disaster (Chapter 7)** followed the King's

Cross fire, which necessarily became a central preoccupation, while simultaneously tackling issues of staff morale and continuing service performance, through the 90+ day inquiry. This period would have been unsupportable, personally, without not only the help of my family, but also the massive support from professional colleagues at home and abroad.

The **Docklands Light Railway (DLR) (Chapter 8)** was an example of *grasping an opportunity and running with it.* Some decisions, rapidly made, have proved to be correct in the long term in a way that I could not have realised at the time. The ultimate outcome of the decision to proceed with the DLR came later with the triumphant performance of transport in London during the Olympics. Also, my small part in winning the bid for the Games confirmed again that *interpersonal* matters can often be as crucial as technical and financial matters when developing a strategy and a deliverable project.

**Eurotunnel (Chapter 9)** is a well-engineered piece of transport infrastructure and provides a good service. But it is not an example of good engineering in the broad sense of my definition. It finished late and cost too much, and the revenue forecasts were disastrous. The client was created much too late and was led by two highly intelligent chairmen, who somehow thought that their role was to act as gladiators fighting the contractors. The contractors were not faultless, but my conviction remains that an intelligent and competent client is *a prerequisite* for a successful project.

After I left the above series of (case study) projects, I **changed perspective (Part III)**, and had time for reflection, while active in academia and the professions, at home and overseas.

**Transport (Chapter 10)**, which was not even regarded as a Civil Engineering subject in the mid-20th century, was my focus when I became an academic at Imperial College — a career bonus.

The presidency of several organisations has not only been part of a contribution to the **Engineering Profession (Chapter 11)**, but it also provided further opportunities to 'spread the word', through ICE and other bodies.

This was followed by professional involvement with **Project Management (Chapter 12)** bodies and collaboration with the actuaries in the study of **Risk.**

**International (Chapter 13)** From the time of my first travel to the USA, I have tried to think (and sometimes live) internationally, academically, professionally and through consulting assignments. The chapter describes some of my contributions.

The **Epilogue** reiterates the role of engineers and the breadth of engineering. It highlights some of the new challenges facing the profession, and describes a 21$^{st}$ century engineer. It concludes with challenges to both educators and employers to give students and young engineers an understanding of the *totality of engineering*. Finally, it encourages them to work more closely together, and with ICE and related professions, thus making it more possible for entrants to make an enhanced contribution to society.

## Bibliography

Cross H (1930), Analysis of continuous frames by distributing fixed-end moments, in *Transactions of ASCE*, Vol. 96, Paper 1793.

Cross H (1952), *Engineers and Ivory Towers*, pp. 55–75, McGraw Hill, New York.

Hindley Sir Clement (1941), Engineering economics, organisation, and aesthetics, in *Journal of the Institution of Civil Engineers*, Vol. 17, No. 1, pp. 49–61.

Ridley T (1954), The aims of an engineering education, Term Paper, University of Durham, Newcastle.

Ridley T (1995), Civil-engineering diversity, First William Fisher Cassie Lecture, in *Proc. of a Colloquium*, pp. 1–12, University of Newcastle, Allenholme Press, Wylam, England.

Inglis Sir Charles (1941), Presidential address, in *Journal of the Institution of Civil Engineers*, Vol. 17, No. 1, pp. 1–18.

# Part I
# The Making of an Engineer

# 2

# Formative Years, Qualification and a New Direction 1933–1965

**Where did I come from and what was my education (in the UK and USA)? How did early site and drawing office experience lead to professional qualification? Why go back to the USA to study a new discipline? How a job offer in Britain led to a return to practice that new discipline?**

## 2.1 The Ridley/Armstrongs

All of my family come from the Northeast of England, where I was born on 10 November 1933 in Castletown, County Durham to Jack (John Edward) and Olive (nee Armstrong).

Jack was born in 1900 at Langley Park, County Durham, the son of John Ridley who left a farm to become first a 'gardener' (domestic) and then a 'horse-keeper' at two collieries. Olive was born at Shotley Bridge, County Durham, in 1905. At that time her father worked as a fitter at Consett Iron Works.

My parents met in Sacriston and married in 1932, when Olive moved to Castletown (now part of Sunderland), where Jack was now a manager for the Wearmouth Coal Co. Jack was a skilled and ambitious mechanical and electrical engineer and, after several years of night-school while working as a fitter, he moved through the ranks until he became an Area Engineer in the National Coal Board.

With each promotion there was a move to a new 'tied' house. We were living on the north side of the River Wear in Sunderland at the outbreak of World War II (WWII). This was 'home' for seven years of my early life though, for much of the time, I was at boarding school.

### 2.1.1 School

Education began at the kindergarten department of Grange School in Sunderland, when I was five. This was followed by preparatory school in Harrogate, Yorkshire in September 1940, still not seven years old, effectively 'evacuated' to an all-boys boarding school, Clifton House. In the first term, I managed to become 4th in the class, but out of only 10, and was top in Arithmetic. By the Christmas Term of 1943 it seemed I was top of the class, when the curriculum now also included English Literature, Algebra, Geometry, Science, Latin, French and General Knowledge.

Next I moved to board at Durham School. It has a long history. One edition of the school register speculated that its origins were in Saxon times. However the only certain date is the setting up in 1414 by Cardinal Langley of a Song School and Grammar School. The School is intimately related to the Cathedral, and so it felt to me. To this day I still get a pleasurable shiver on hearing the sound of a choir echoing high in the roof of a Cathedral.

I started well at Durham — with high marks and several subjects ranked as 'very good' or 'excellent'. Luce, the headmaster, simply said, 'very promising'. My housemaster, Hall, told my parents that, 'providing nothing happens to hinder him, I prophesy that he will, in the course of time, be one of the best leaders we have had here.' My sporting prowess was limited.

### 2.1.2 University

When I left school it was decided that I should study Civil Engineering at Kings College, Newcastle, in those days part of Durham University, and that I should live at home, after 12 years at boarding school.

In the early 1950s, an undergraduate degree in Civil Engineering took three years. The first year consisted of maths and science. Entrants who had already achieved A-levels in these subjects could enter the second year, and complete the requirements for a pass degree in two years. The amount of knowledge and understanding absorbed was vastly less than we imparted to Imperial students in four-year course when I became Professor, and subsequently Head of Department, 40 years later.

It was still necessary to study three years in total to obtain a degree. So those who had entered second year stayed on and gained an Honours degree. To do this, each had to carry out a project. I spent much of third year, in addition to some Economics, German and more Mathematics on my thesis, 'A Study of Rectangular Concrete Members in Torsion' (Ridley T, 1955). That, and my earlier work, gained a 2:1 Honours degree.


## 2.2 USA

I then fulfilled an ambition to go to the States. From August 1955 until May 1957 I was in America, having won a Fulbright travel scholarship. First it was necessary to find a University where I could study and I settled on Northwestern University in Illinois, whose Civil Engineering Department was offering research assistantships, requiring 5 hours of assistance per week, which allowed me to study for a Master of Science degree in one year. I applied and was accepted, both working and studying at the same time over 12 months.

The research programme addressed the static and fatigue properties of high-strength low-alloy steel joints related to those of carbon and silicon steel joints (Ridley T, 1956a).

The programme of study indicated the greater *breadth* of an American degree. The main focus of my undergraduate degree at Newcastle had been on Structures, so I opted to concentrate on Soil Mechanics at the Master's level. However, it was also open to take a number of courses outside Civil Engineering. Thus, I chose three nuclear engineering courses — Principles, Processes and Design and, with two American colleagues, submitted a course project, 'A Preliminary Design of a Nuclear Power reactor for a Radar Warning Station' (Ridley T *et al.*, 1956).

I was also very interested in management subjects and thus studied the 'Economics of Technical Change', and wrote a paper on 'Nuclear vs. Conventional Power' (Ridley T, 1956b), again inadvertently helping my future employment and broadening my interest in engineering. To this was added Industrial Organisation and Management, and the Business, Legal and Ethical Phases of Engineering.

## 2.3  Go West Young Man

At the end of the year at Northwestern, under the terms of the Fulbright scholarship, it was possible to stay and work for a further year. My Professor wrote to Shannon and Wilson in Seattle who offered me a job at $450 per month, an improvement to my standard of living, and going to the northwest of the US seemed like a good idea.

Bill Shannon and Stan Wilson had both been students of the great Terzaghi at Harvard. Born in Prague and having worked in Austria, Istanbul and at MIT, he eventually moved permanently to the USA when he joined the faculty at Harvard, becoming known as the father of Soil Mechanics — or Geotechnical Engineering as it is known today.

The first role in the S&W office was as a draughtsman and technician, carrying out calculations, under instruction, relating to slope stability, very necessary for apartment foundations in Seattle, and missile sites in Montana. I had one particular advantage in the firm. Being a bachelor, I could easily be sent out on jobs, thus allowing American employees to stay at home with their families. This suited me very well as it gave me travel, experience and responsibility at the age of 23.

The first site job took me over the Cascades (mountains) to Chelan in the Columbia River Valley. The Columbia was a favoured river for dams for hydroelectric power, the Grand Coulee being one of the most famous. Wells Dam, long since constructed, was on the list of future projects. The task was to meet up with a three-man drilling crew that the firm had appointed, to supervise a site investigation and take and record samples as the drillers produced them. The samples were then shipped back to Seattle for analysis.

I was there in December, when the water levels were at their lowest, in order to get as far into the riverbed as possible. Being December the weather was perishing, and it was a necessary precaution to buy long-johns to help to keep warm.

It was in Chelan that I first experienced 'there's more to engineering'. Having just returned from meeting the drill crew on site, and arranged to meet again the following morning to start work, I had only just checked into a motel for a month and was still unpacking, when the telephone rang and an American voice introduced himself as a local attorney. He represented Henry Miller, a Native American–Indian and wanted me to know that I had been trespassing on Mr. Miller's land. He then gave me a lesson in American history, in particular how badly the white man had treated the Indians.

I apologised and explained my position, and said I would get back to him as soon as possible. I rang the Headquarters and after they had spoken to the attorney, we started work as planned. Three days later, however, Henry emerged from his house, on site, as it had been arranged that he be employed as a labourer for the drill crew. The arrangement during the ensuing month was very amicable, but possibly the big fish for the attorney, and Henry, was to be a major compensation package for his land if and when the dam went ahead, as it did.

The next trip was to Arcata in Northern California, close by Eureka. A pulp mill was to be built on a sandspit by the ocean. It was therefore necessary to carry out a series of measurements of the sand density in order to calculate the degree of settlement that might arise from the effect of vibrating machinery.

The final trip was much further away from Seattle. The Republic Steel Co. had a plant in Cleveland, Ohio alongside the Cuyahoga river, subsequently renowned for being so polluted that it became 'the river that caught fire'. The piling that held the iron ore dock in place had started to move, and it had become inoperable. Republic had called in Terzaghi himself to advise and, after a day's visit and analysis, he suggested that they get in touch with Stan Wilson to fix the problem. This was very often Terzaghi's style. He would understand the issues, charge a fee for his analysis and advice, but not hang around for the next steps.

Stan Wilson had decided that he needed to know more about the ground movement and that he would use a new tool that he was just developing, a 'slope indicator'. This was used to monitor lateral ground movements, with casing installed in a vertical borehole that passed through suspected zones of ground movement into stable ground. The casing was special purpose grooved pipe. It had three purposes. It provided access for the slope indicator probe, allowing it to obtain subsurface measurements; it deformed with movement of the adjacent ground; and it controlled the orientation of the probe.

My job, with a crew, was to supervise the drilling of a vertical borehole, into which the casing was installed, and then to obtain the initial survey figures, sending them back to Seattle. In all this took four weeks. Soon it was time to return home after two wonderful stretching years in USA.

## 2.4 Return Home

### 2.4.1 Site Experience

Being involved with whatever was modern and 'the latest' has always been fascinating. When considering what to do on my return home, the new civil nuclear power programme seemed exciting. The government had recently encouraged the creation of four contractor consortia to build new civil power stations.

Following the philosophy, 'never be afraid to ask, the worst they can say is no', I wrote to Sir Claude Gibb, Chairman of the Nuclear Power Plant Company (NPPC), a diminutive but highly dynamic Australian industrialist. He wrote back that there was nothing he could do for me at a distance but I could go to see him on return home, which I did. After a stimulating hour he said, "we will appoint you to the Resident Engineers' staff at Bradwell Power Station. Go to Knutsford (in Cheshire) to meet Joe Sayers, who is the Chief Executive of the consortium, where you will be briefed". I spent a month in Cheshire, and then went to Essex.

Following the successful opening of Calder Hall in Cumbria, the world's first industrial-scale nuclear power station, construction began at Bradwell-on-Sea in Essex in January 1957, on the edge of the Blackwater Estuary, 50 miles from London. At about the same time AEI-John Thompson began to build a similar station at Berkeley in Gloucestershire.

Bradwell was a 300 megawatt station, using two graphite-moderated, gas-cooled reactors. The boss, Andy Young, had been responsible for the construction of Calder Hall. He had a mixed team on his RE staff. I, together with all of the mechanical and electrical engineers, was paid by NPPC. But I was unique in so far as civil engineers were concerned, who were all on the payroll of Sir Robert McAlpine, the civil contractor.

Site life for a bachelor was very comfortable. NPPC provided accommodation for its staff in the Ranch, a new temporary building with bathrooms and showers, and a restaurant and bar that provided excellent meals.

Social life included the village Dramatic Society where I played in Laburnham Grove by J B Priestley, and produced the Society's next performance, Alan Melville's comedy, 'Simon and Laura'. A mini-crisis arose when NPPC were bidding to build Italy's first nuclear power station at Latina, near Rome. A member of the cast, Brian Sanders, was required to participate in the tender preparation in Knutsford. There was no option but to play his role myself. The local newspaper, Maldon and Burnham Standard was impressed, and said that there was something of the master touch in the way this subtle comedy about human nature was handled by the producer, Tony Ridley, who also played the important role of David Prentice, a TV producer. I have no doubt that both debating and theatre have given me life enhancing skills.

While working at Bradwell, I married Jane Dickinson at Longhoughton church, Northumberland, after which we lived in an idyllic cottage in Whickham Bishops for four months. I had by now discussed my future with Chief Civil Engineer, Houghton Brown, in the Knutsford headquarters of NPPC. Having had site experience, it was necessary to work in the design office towards my professional qualification. In December 1959, as newly-weds, we moved north to a ground floor flat in Altrincham, Cheshire to begin working in the design office.

## 2.4.2 Design Experience

The Nuclear Power Group was formed on 1 January 1960, a partnership between NPPC and AEI-John Thompson. NPPC had been awarded the Bradwell contract by the Central Electric Generating Board (CEGB) in December 1956 at the same time as the Berkeley contract. While I was at Knutsford, both went critical on virtually the same date.

In June 1960, TNPG were awarded the contract for Dungeness 'A' Power Station in Kent, after which arrangements were made for the construction of a hydraulic model of the cooling water pump house, in accordance with the contract. The CEGB required that 'the design of the pump suction chamber should be determined by the pump manufacturer and arranged to give the best possible hydraulic conditions, consistent with the minimum depth of excavation necessary to obtain satisfactory operating conditions'. The pump manufacturer was Drysdale and Co. Ltd.

It was the responsibility of McAlpines, as designers of the civil engineering works on behalf of TNPG, to verify that the flow from the inlet to the pump house as far as the suction docks was satisfactory. The basic layout of the pump house was arrived at on the assumption that satisfactory hydraulic conditions could be obtained in a very compact design.

It was necessary to investigate the hydraulic performance of the pump house and, if required, to carry out modifications. The main object of Civil Department's tests was to show that the conditions in the fore-bay and screen areas were satisfactory, and that the flow to each pump bell-mouth did not vary beyond acceptable limits. The object of Drysdale's tests was to enable them to guarantee that the conditions in the suction docks, under all reasonable operating conditions, would be satisfactory.

As this work fell within the responsibility of Civil Engineering Department, I was appointed to design and supervise the construction of the test rig, at the firm's Knutsford headquarters, then to carry out flow tests, and write up the results. The initial model test results, with the proposed design, indicated unsatisfactory suction dock flow conditions at the lower water levels for certain operating conditions. However, development tests retaining the basic pump house design were successful in improving the flow conditions and reducing swirling in the suction dock considerably by streamlining both normal and cross flows (Ridley T, 1962a).

Apart from the particular work on the hydraulic model I was one of the junior engineers living behind one of a row of drawing boards in the drawing office. Roger Hyde, Deputy Chief Civil Engineer (a mentor) took me under his wing. He subsequently became the Chairman of Allott & Lomax. He helped me greatly and is one of the people to whom I feel a debt of gratitude.

Having worked on schemes and layouts for the cooling water tunnels for the tender for Sizewell nuclear power station, detailed design for the Latina contract, as well as design work for the Oldbury station tender, I then spent my time in Knutsford on the Dungeness 'A' nuclear power station contract — the Syphon Recovery Chamber for the CW works and the hydraulic model tests described above.

## 2.5 Professional Qualification

At about that time I began to teach part-time. I had previously refreshed my elementary structural engineering by correspondence course while in Bradwell. Armed with that improved knowledge I started teaching Theory of Structures to third year Ordinary National Certificate students at Peel Park Technical College in 1960, and then at Salford Technical College in 1961–1962.

Drawings and calculations for my ICE professional interview were dated 8 February 1961. The submission included a resume of Training and

Experience covering my degree, Honours thesis, vacation work, Fulbright scholarship, research and MS at Northwestern, work to date for TNPG, and part-time lecturing at Salford Technical College (Ridley T, 1961b). Drawings of the reactor building at Bradwell, works at Latina, and the Dungeness recovery chamber, with associated calculations and bills of quantities were also submitted. Together these were parcelled and dispatched to ICE for an AMICE — the equivalent of MICE today.

In addition, a professional interview at One Great George Street, the ICE headquarters, was required. London was slightly familiar but it was all rather forbidding, not least because the panel of three was chaired by Charles Haswell, one of the youngest partners in Halcrow, who later set up his own consulting firm.

Halcrow had been the CEGB's engineers for Bradwell power station but, because the design of the works was in the hands of Sir Robert McAlpine, they had little opportunity to intervene to any significant effect. Haswell in particular, did not like this arrangement and he was very challenging regarding the fact that the civil resident engineers were on McAlpine's payroll. I was profoundly relieved to be able to say that no, I had not been on the payroll of the contractor on site, but of NPPC. Evidently all was forgiven, for I passed my AMICE and became qualified. Indeed, later in Hong Kong, Haswells were our tunnelling consultant and Charles became a friend, who used to come for dinner whenever he visited the project.

## 2.6 University of California

During the work for my AMICE, I decided that nuclear power was not as interesting as I had hoped. (Jane says that I had always intended to return to the States.) I wondered about becoming an academic and decided to study for a PhD in the new field of Transport. Happily, seeking to become a transport analyst, planner and engineer, I already had experience behind me, albeit at a junior level, of 'real civil engineering' — concrete, steel, geotechnics and hydraulics, with 'muddy boots' thrown in.

### 2.6.1 A Significant Conference

In April 1961, Tom Williams ran a conference at King's College on 'Urban Survival and Traffic' (Williams, 1962) which I attended. Tom was the Rees

Jeffreys Reader in Highway and Traffic Engineering. The organising committee — Professors Fisher Cassie (Civil Engineering), Napper (Architecture) and Allen (Town Planning) described the purpose of the conference as an important first step towards civil engineers, architects and town planners working together to face the 'social, economic and visual problems created by traffic in cities'.

This important conference inspired me to follow a new intellectual and professional life. Among the speakers were (later Sir) Colin Buchanan from the Ministry of Transport; academics from Northwestern University's transport programme; Alan Voorhees, represented the Automotive Safety Foundation in Washington DC; Reuben Smeed from the (then) Road Research Laboratory whose Committee, including (Sir) Christopher Foster, Michael Beesley and Gabriel Roth, 'invented' congestion charging; and Wilfred Burns, City Planning Officer of Newcastle. Most significant for me was the presence of Professor Norman Kennedy of the Institute of Transportation and Traffic Engineering (ITTE), at the University of California, Berkeley where I subsequently studied.

## 2.6.2 Why Go Back to the States?

It was possible to study for a PhD in Transport in the UK, but it was not a well-developed subject. It was also clear that being a poor student in warm California was preferable to being a poor student in cold Britain. Furthermore, since I was starting a new subject from scratch, an American PhD seemed preferable. As compared with a minimum of three years of full-time research in Britain, the American PhD included two years of post-graduate study and a research project of only one year's duration. This seemed a better grounding for me.

Having already benefited from a Fulbright scholarship, it was unlikely that another scholarship would be available. I had supported myself via a research project at Northwestern. Might I do the same again — and where?

A meeting with Professor Kennedy at the abovementioned Conference, was fortuitous as he gave me an offer of a place at the ITTE in Berkeley, California, with financial support. Now it was urgent to complete my AMICE. Autumn 1962 was a good time to start my studies. Whether by coincidence or not, Jane became pregnant. Work for ICE was completed and the professional exam passed. Professor Cassie wrote, expressing his disappointment that I was leaving the country but agreeing to act as referee.

The decision to return to the States to study for a PhD had been viewed by family and friends as some kind of madness. We both had good jobs and a young baby. I was 28 and Jane 26, and most of our friends were 'settling down', getting mortgages and generally getting on living and working in the UK. At the time it seemed like an easy and exciting choice to us — even if our families were both sad and apprehensive for us. Looking back we have absolutely no regrets.

### 2.6.3  First Year at ITTE

Professor Kennedy (another mentor) took us to see our new home in Albany. The two-storey buildings on the site accommodated some 900 graduate students in ex-Navy housing brought to California from Oregon. We knew that our upstairs apartment would suit us and our daughter very well, but he was appalled to discover what simple accommodation he had arranged for us to live in. The family was supported by my working 20 hours per week on research for the Department, financed by the California State Highway Department.

The PhD programme of the Institute would take three or four years at least. A PhD required two years of study and a further year to complete a dissertation, and required study and examination in three subjects — a major (Transportation) and two minor subjects (I chose Statistics and City Planning). At the end of the second year I would be examined by a Board, representing the three subjects. If I passed I would qualify to proceed to my dissertation.

The first year's programme consisted of basic study of Transportation Engineering, including sections on Traffic Flow Theory and Network Analysis by Professor Bob Oliver, who was to become my dissertation adviser; Theory of Probability; Queuing Theory; City and Metropolitan Planning for Engineers. It was also possible to 'audit' (i.e. sit-in without examination) additional classes and I took courses on Economics, and on Networks by D R Fulkerson of the Rand Corporation.

### 2.6.4  A Formative Paper

For a course on 'Analysis of Transportation Systems', I developed a term paper on the '1956 Federal Highway Act and Urban Areas' (Ridley T, 1962b).

Already an urbanist, I went to the US in the mid-1950s partly because the country and its political processes were fascinating. The paper has remained relevant, because it was an historical review of the development of major highways in the States, and it reinforced a developing interest in transport policy and the breadth and scope of engineering. It began by stating that the politics of the country were extremely complex and suggested that the 1956 Federal Aid Highway Act was a compromise between competing interests of citizens and pressure groups.

It also chronicled the history of relevant legislation from the 1916 Federal Aid Highway Act that represented central government's first major commitment to a road programme. The 1956 Highway Act provided for a 13-year programme of improvement of 41,000 miles of the Interstate System. The Highway Revenue Act established a highway Trust Fund in the Treasury into which user-taxes were to be paid. Gas tax was increased and a stipulation was made that the programme would be on a pay-as-you-go basis. Crucially, the system was named the 'National System of Interstate and Defence Highways'. That played a big part in convincing the reluctant of the need for the system.

### 2.6.5 Second Year

In the Fall term of the second year graduate-level courses were in Theory of Probability; Industrial Engineering, which would have been better known as Operations Research; and Statistics and City Planning in Spring 1964. My study included lectures on Linear Programming by Prof Dantzig, who had invented the subject while doing research during WW II, an experience I can only describe as 'challenging'. In September 1963, I inquired of Colin Buchanan, then the Advisor on Urban Road Planning to the Ministry of Transport in London, if I could have access to preliminary, non-confidential material from his upcoming report to the Minister. He suggested that I should wait for the publication of the Report but thanked me for the good wishes for his new Chair at Imperial College — (a position I followed him into 28 years later).

It was at about this time that a realisation about PhDs began to dawn on me. No doubt some geniuses gain PhDs and go on to win Nobel Prizes, but for most of us, certainly for me, it was quite different. You don't achieve a

PhD without some basic intelligence but getting a PhD, is 10 percent inspiration and 90 percent perseverance — my mother's favourite word.

### 2.6.6 Teaching

As a result of Bob Oliver deciding to return to work full time in the Industrial Engineering Department, I was asked to teach that section of the basic graduate course, which he had taught me. It covered — Car Following Theory, Fluid Flow Theory, Simple Intersection Models and Elementary Network Flow Theory. I was appointed Acting Assistant Professor (the lowest rung on the US academic ladder) and paid by the University.

That suited me well, since I intended to become a Lecturer in Transport when I went home. Coincidentally, (now Professor) Vuchic, a Serbian–American who became renowned at the University of Pennsylvania in Philadelphia, was briefly one of my 'students'.

I also taught practising engineers. ITTE participated in programmes put on by the Engineering Extension of the University of California, and Wolf Homburger and I gave a series of lectures in both California and Nevada. A two-day programme in Sacramento was titled 'Introduction to Traffic Estimation'.

> "This course will present a resume of trip estimation procedures, including land use inventories, trip generation studies, trip distribution and assignment theories and the preparation and testing of transportation network plans. Some reference will be made to computer applications; an elementary knowledge of computers, therefore, would be helpful to those attending the course, but is not a prerequisite."

### 2.6.7 Dissertation

The topic I chose for my dissertation was a network problem — investment in a transport network.

Bob Oliver, who became my Advisor, not being an economist, no '£' sign or even a '$' sign, appeared in the work. To Bob, the project was simply the solution of an applied mathematics problem and, furthermore, he was interested in mathematical elegance rather than tackling a real life issue. Nonetheless, when it came to the point, he gave me a great deal of help with the proper presentation of the mathematical analysis.

The title of the dissertation (Ridley T, 1965) was, 'An Investment Policy to Reduce the Travel time in a Transport Network'. It is best described by the Abstract.

"The principal aim of a transport study is to produce a plan for a future transport network. This network should satisfy the traffic demands placed upon it and give service on the basis of some acceptable criteria within certain budgetary, political and social constraints. Out of a large number of engineering design problems that naturally arise, this dissertation considers a special problem of economic investment in a transport network."

The nature of the dissertation highlights the difference between an American and a UK PhD, which would have been entirely research based. Of course, that was precisely why it was so useful for me. The two years of taught courses gave me what I needed for a career in an emerging area of Civil Engineering. Whether it would have been adequate for an academic career is not clear. Also the 'solution' for the problem identified was definitely a theo-retical one. It in no way addressed the totality of real life. Nonetheless, as with all research, it built on previous knowledge and took a step forward in the direction of a solution to the ultimate problem.

Another learning opportunity arose at a symposium in New York in June 1965 on the Theory of Traffic Flow. My attendance was funded and a paper was accepted. (Ridley T, 1967). This was a special case of the problem on which I had embarked for my dissertation, and the Symposium gave me a brief insight into the work of the great men of the subject — Leutzbach from Karlsruhe in Germany; the Nobel Prize winner Ilya Prigogine; and Reuben Smeed and Richard Allsop from the UK. Richard, much later, became a colleague when I moved into academic life in 1991.

## 2.7 Settling on a Future Direction

As usual, when thinking about future directions, I sought advice from Professor Cassie. He had suggested before I left the UK that I should get some teaching experience in the USA. I explained I had much enjoyed the time in the States — as much in City Planning, Queuing Theory and Linear Programming as in Transport at ITTE, which should assist in broadening Transport study in the UK.

He had no doubt what I should do. He had spoken to people he knew at the Road Research Laboratory (now TRL). My best plan would be to come home now (by which I assumed he meant when I had graduated) for 'things are becoming very vigorous and you would do well to get in on the ground floor'. What the correspondence with Cassie, and others, did illustrate was that my decision to go to Berkeley to educate myself in a new field had been very timely and would pay off in a way that I had never contemplated.

Then a completely unexpected opportunity with the GLC appeared, as is described in the next chapter, and I was being pressed to get back to London. But there was a problem with producing an 'optimal' solution to the challenge of the PhD dissertation. The first 'solution' did not pass the test as I had discovered a simple counter-example (if just one example exists, however trivial, where a solution is demonstrably not optimal then, by definition, the general solution cannot be optimal). I frankly did not know what to do next. Bob Oliver was friendly, but not reassuring. If the problem could not be solved the dissertation would be incomplete. I would not graduate with a PhD, after three years of very hard work. Alternatively, the option was to stay on at ITTE to finish the dissertation, and lose a very attractive job offer in London which I had accepted.

After further delving in the literature, I did develop a satisfactory solution and completed the dissertation, meanwhile being bombarded by urgent communications from Peter Stott and Brian Martin at the GLC. There was a new Department to set up and they wanted the No. 2 in post ASAP.

On returning home there were two further opportunities to publish papers based on my dissertation. The first was in Transportation Research (Ridley T, 1968) and an invited paper to the annual conference of the British section of the Regional Science Association was also published (Ridley T, 1969).

## Bibliography

Ridley T (1955), A study of rectangular concrete members in torsion, Honours Thesis, University of Durham (Newcastle).

Ridley T (1956a), Static and fatigue properties of high-strength low-ally butt joints related to those of carbon and of silicon butt joints, Research Council on Riveted and Bolted Structural Joints, Project VI, Department of Civil Engineering, Northwestern University, Illinois, USA.

Ridley T (1956b), The economics of technological change: Nuclear versus conventional power, Term Paper, Northwestern University, Illinois.

Ridley T (1961b), AMICE report, Knutsford, Cheshire.

Ridley T (1962a), Report on the model test of the circulating water pump-house at Dungeness Nuclear Power Station, TNPG 268, Knutsford, Cheshire.

Ridley T (1962b), 1956 Federal Aid Highway Act and urban areas, Graduate Report, ITTE, Berkeley, USA.

Ridley T (1965), An investment policy to reduce the travel time in a transportation network (PhD dissertation), Operations Research Center (ORC 65-34), University of California, Berkeley, USA.

Ridley T (1967), Investment in a network to reduce the length of the shortest route, in *Vehicular Traffic Science, Proc. of the 3rd International Symposium on the Theory of Traffic Flow*, pp. 235–236, New York, Elsevier.

Ridley T (1968), An investment policy to reduce the travel time in a transportation network, *Transportation Research*, Vol. 2, pp. 409–424, Pergamon Press.

Ridley T (1969), Reducing the travel time in a transport network, in *Studies in Regional Science*, pp. 73–87, London, Pion.

Ridley T *et al.* (1956), Preliminary design of a nuclear power reactor for a radar warning station, Term Paper, Northwestern University, Illinois.

Williams T (1962), Urban survival and traffic, *Symposium Proceedings*, London, Spon.

# Part II
# Practising Engineering 1965–1990

# 3

# Greater London Council 1965–1969

How did this first managerial job provide the opportunity to develop a new discipline with a young and keen team, who all went on to interesting careers? What were the shortcomings of the ground-breaking London Transportation Study?

Some lessons

- If you apply for, but fail to achieve, a top job, you may find another to your liking.
- Try to discuss potential career changes with your spouse/partner.
- Learn to live with publicity.
- Early examples of transport and politics.

## 3.1 The New GLC

In January 1965 an advertisement appeared in the ICE Proceedings, for posts in transport in the new Greater London Council, formed on 1 April. It read,

> "The Director of Highways and Transportation (PF Stott MA MICE), is assembling an organisation (which will be unique in local government) to be responsible to the GLC for research into and development of a comprehensive transportation scheme for an area of 600 square miles and 8 million population; planning the future road system, including urban motorways; design and construction of the main thoroughfares and intersections; traffic management schemes throughout the metropolis."

The structure of the new Department was to have two Chief Engineers — Highways and Traffic and a Chief Advisor on Transportation Policy, beneath who would be Assistant Chief Engineers of Highways Planning, Highways Construction, and Research, and a Traffic Manager. Below that level there were to be a number of Divisional Engineer positions. The ACE (Research) was described as 'a person who combines excellence in the practical sciences with resource and imagination and the ability to lead and inspire a mixed team, including mathematicians, traffic engineers and electronic engineers, in the pursuit of land-use/transportation studies and in the solution of traffic, planning and constructional problems thrown up by the activity of the department'.

This was interesting so I wrote to the Clerk to the Council with my CV and the following statement:

> "I see the task of the Assistant Chief Engineer (Research) as being responsible for mobilising the resources within the organisation, through consultants, the universities and the Road Research Laboratory to the solution of real problems of transportation planning and construction. I strongly believe it to be imperative in Britain today to encourage basic research, but more importantly to bring the results of such research to bear as quickly as possible on contemporary problems. My earlier broad experience and my recent specialised knowledge may make me the person to do this within the Department of Highways and Transportation."

London had been governed since 1889 by the London County Council (LCC), well regarded as a good example of metropolitan government and where Herbert Morrison, subsequently a member of the post-war Labour Cabinet, made his name. It covered what are now the boroughs of Camden, Greenwich, Hackney, Hammersmith and Fulham, Islington, Kensington and Chelsea, Lambeth, Lewisham, Southwark, Tower Hamlets, Wandsworth and Westminster. By 1963 the London urban area had expanded into the Home Counties and the Greater London Council was created.

The advert whet my appetite. They were looking for an ACE (Research). Why not? All the required skills were exactly those that I had been acquiring at Berkeley and there were other jobs, reporting to the ACE (Research), also available. I applied, but nothing happened for several weeks. Then, on the same day, I received a letter and a telephone call. The letter, from the GLC, said that an ACE (Research) had been appointed, but was I interested in any other jobs?

Lesson — If you try for, but fail to achieve, a top job, you may find another to your liking.

## 3.2 An Opportunity

The telephone call came from an Englishman at Harvard — Brian Martin, later the M of MVA, and a former first-class honours student at Imperial College. He said that he had been appointed to the top job and, if I was interested in other posts, the GLC wanted me to fly to Boston to meet him. I was interested, but only had a narrow window between the end of the teaching term and supervising examinations. We arranged dates for our meeting and I bought air tickets. Just two days before I was to fly, another telephone call. The GLC wanted me to fly on to London for interview, after I had met Brian.

This was just possible in a five-day window, so I flew by helicopter across the Bay to San Francisco and then on by plane to Boston. Brian met me at the airport. I was to have dinner with him and his wife, Susan, and stay overnight with them. The interview would be on the next day. We spent Saturday talking about our respective histories — rather short as we were both young — and Brian telling me how he saw the task for the GLC. The 'interview' complete, he dictated his reference to Susan which, when typed, was sealed in an envelope and carried by me on the overnight plane to Heathrow and subsequently given to Stott.

Already jet-lagged, and now eight clock hours from California, I was not in good shape after a poor night's sleep on arrival in London for a meeting with Peter Stott. Brian's letter must have been complimentary as, after a discussion about my background, Stott said that he would like me to be interviewed by the Members' Panel who would decide whether to appoint me to the position of Divisional Engineer, effectively No. 2 to Brian.

I had the interview, which went well and, after some time waiting outside the interview room, I was called back to be offered the job and invited to go back to California to discuss with my wife whether I wanted to take it.

Lesson — Try to discuss potential career changes with your spouse/partner.

Wow, I thought, this was how things are done in the big world that I now inhabited. Indeed, the experience led me to expect something similar when I later went for interview for the job of Director General of the PTE in Newcastle, as I shall recount. I could have not been more mistaken on that occasion.

There was no doubt that I wanted to take the job, and all thoughts of an academic career were soon dismissed. The opportunity was too good to miss — new job, in a new field of transport, with a substantial salary for my age, and in the heart of London. Jane shared my opinion and I soon communicated the decision to Stott back in London. The main issue was, how soon could I start? The dissertation had to be completed first. I began to receive increasingly anxious telegrams and phone calls pressing me to get to London as soon as possible because, of course, Stott and Brian with others were eager to create the team within the Department of Highways and Transportation.

Eventually we left Berkeley, said goodbyes, and flew to London with Sarah and Jon. There was one casualty. I had long promised Jane, as a thank you for her support during three years, that we would not fly direct but go via Mexico City for a sight-seeing and relaxing holiday. It never happened and it was years later before I was eventually able to take her to Mexico.

After an overnight flight we checked into an airport hotel, then to the Richmond Hill Hotel for five days, finally to a top floor apartment in Friars Stile Road, Richmond for two months, both of which must have whet our appetite for a return to Richmond 15 years later. We eventually settled on renting a tiny lodge in Surbiton. It was a beautiful sunny summer, ideal for spending long days in the garden with our two small children and our new baby Michael, who was born in Kingston hospital.

### 3.2.1 Early Days

The first task was to forecast the traffic demand and parking requirements for public and trade exhibitions at a potential National Exhibition Centre at Crystal Palace. In the event no such Centre was built in London, Rather the National Exhibition Centre was built in the Midlands.

With a PhD, and a variety of practical Civil Engineering experiences — pre-stressed concrete, riveted and bolted structural joints, a hydraulic model test, and geotechnical consultancy — but no actual experience of traffic forecasting I was fortunate to have a senior lieutenant, Paul Prestwood Smith, who had. We garnered such information as was available from the London Traffic Survey, together with data from other Exhibition Centres. Our report (Ridley T *et al.*, 1966) made it clear that Crystal Palace would not be an ideal location.

Another specific task arose after the election, when the Tories took over from Labour. The Tories, in their election manifesto, had proposed a ten-point plan. The tenth item was for the construction of a monorail system in Central London. The proposal was developed by Brian Waters (1967), an Architect, in 'Get Our Cities Moving'. The Tory leader, Desmond Plummer said in a Foreword, which noted the congestion in Central London, that Waters had suggested a solution which might well overcome this apparently insoluble problem. His ingenious and exciting adaptation of the monorail concept for the central and congested area of London deserved the closest study.

My task was to carry out, 'a preliminary assessment of the feasibility of monorails, for passenger distribution in Central London.' The report (Ridley T, 1967) addressed the system, description, route, transport characteristics, physical feasibility, effect on the environment, capital costs, estimates of traffic demand, and operating policy. It concluded that, the imposition of an overhead system would pose major environmental problems and, even if major transport benefits were to accrue, the effect on the environment might not be acceptable to the public. Two photographs in the report showed an elevated monorail travelling along Fleet Street and up Regent Street, which meant that the idea was 'dead in the water'.

Other responsibilities included the London Transportation Study, the creation and development of a traffic section, and the conduct of a series of studies requiring the collection and analysis of traffic data for the Department. First we had to recruit staff; Stott had already made a good start. Before Brian and I had arrived, he had already recruited some 10 people whom he sent on a GLC sponsored course at Birmingham University under Professor Kolbuszewski.

Brian and I attracted attention and became a small issue in the Council election. One newspaper ran a piece saying that the GLC had become such a powerful magnet for traffic experts that they had succeeded in luring back a considerable number of brains in the throes of absconding across the Atlantic. One of these, Brian Martin, the GLC's Assistant Chief Engineer, had gone to MIT with a first class honours degree from Imperial College and

**PHOTO 3.1.  No Monorail in London.**

was Assistant Director of Transport Research at Harvard when the GLC booked him by transatlantic telephone. Tony Ridley, In Charge of Transport Planning, had been tempted back from the University of California.
Lesson — Learn to live with publicity.

### 3.2.2 The New Transport Planners

Brian and I recruited a significant number of people and thus developed considerable experience of the interview process. 'Traffic' was led by Ken Huddart, who became one of the leading lights in the country in his specialisation. Among those we hired were David Bayliss and Tony May, who subsequently became Professor at Leeds University and also Deputy Vice-Chancellor. Malcolm Buchanan, son of Colin, joined us when he came out of the Army. He later left to run the successful Buchanan consultancy, following his father.

There is no doubt that, transport analysis and planning, being a new specialisation in Britain, benefitted significantly from the resource and talent put into it, started in the mid-1960s by the Department of Transport, and led by Chris Foster when he was working for Barbara Castle as Minister. Among the names in his team were Alan Wilson, David Quarmby, Stewart Joy and others. Equally the GLC team made an important contribution.

The consultants for the London Transportation Study were Freeman Fox and Wilbur Smith and the FF partner responsible was Oleg Kerensky, a Russian. His father, Alexander, was Prime Minister of Russia July–November 1917 and, after the Bolshevik revolution, he emigrated to Paris and then moved to the USA. Oleg studied civil engineering and made his career in the UK.

### 3.2.3 Transport Planning Lessons

LTS was the first comprehensive transport study in the UK that used modern methods of analysis, following Pittsburg and Chicago in the USA. There were two early reports, on the prior London Traffic Survey (GLC, 1964 — Vol. 1, Existing traffic and travel characteristics in Greater London) and then (GLC, 1966 — Vol. 2, Future traffic and travel characteristics in Greater London). They suffered from one major difficulty — the complexity of computer methodology they required.

I wonder if my generation would have made a more effective contribution if we had carried out our planning without the benefit of the computer. Too often figures emerged from the computer so late that we did not have proper time to interpret the numbers. In addition, the mathematical models lacked a capacity constraint function. In layman's terms, the traffic flows that emerged were what would have been correct if roads could have been built as wide as necessary. This was clearly nonsense, and was understood to be so. But very high forecast flows did mislead the highway planning thinking.

I expressed my general concerns about the current status of transport analysis and forecasting in contributing to the discussion at a Transportation Engineering Conference in London (ICE, 1968).

"Transport planning is a co-operative venture. A number of different specialists is employed — engineers, mathematicians, economists, and many other people. One of the things that the studies and mathematical models have done is to provide a common language in which people with very different backgrounds can discuss the transport problem. But, one of the greatest shortcomings of the studies, as at present constituted, is a proper interpretation of the results. One of the problems is that they are so complex that it is difficult to understand what their proper significance is. Research into new techniques must continue, but we will only get the correct balance in the different phases of our work if we constantly ask the question: how will the decisions be altered if the techniques are changed?"

The network plan the highway engineers asked us to test consisted of four Ringways. Only the fourth, approximately the M25, ever came to fruition. There was a rationale for the ringways in that the DoT planners were developing a series of radial roads pointing at the heart of London. GLC had to find a way of distributing this traffic around London, and that thinking went into the Greater London Development Plan.

In 1967, Brian Martin assumed wider responsibilities and I became Chief Research Officer which continued until I left to become DG at Tyne and Wear Passenger Transport Authority. Although there was little which we had changed on the ground in London, we did leave some significant papers on the subject of transport in London. 'Movement in London' (Ridley T *et al.*, 1969) described the transport research studies and their context. 'The London Transportation Study and beyond' (Ridley T *et al.*, 1970) concluded,

"No network which seems feasible, either financially or politically, would meet in full the potential demand for movement by road; whatever investment is made in the road system a substantial degree of control of movement will be advantageous; opportunities for the improvement of public transport by major new links exist, but they are limited in number; benefits are likely to result from improvements in the existing rail system and improving the operation of buses; the overall use of public transport is not expected to fall, given the expected levels of investment in roads and public transport on the assumption that, with control of movement by road, car owners will still be prepared to make their trips by public transport; because car ownership is increasing very rapidly, even with substantial investment in roads, there will not be any increase in the freedom of an individual car owner to use his car; furthermore, if investment in roads is severely curtailed the degree of control of the use of the car which would be necessary is much more severe than the public has yet contemplated and might be prepared to accept.' Even in retrospect these were valid conclusions, except the reference to the limited number of opportunities for new public transport links, and even that did not become clear until about 1985."

The work of the transport planning and research team had been ill-fated from the outset. In particular the concept that, 'transport is politics' became obvious. While a student in California I had observed the 'freeway revolt', in San Francisco, but it never seemed possible that such a revolt could happen in London. The 'Homes Before Roads' movement became very powerful. Furthermore, Peter Stott had made an unhappy decision at the very outset of the GLC. In response to a public relations request for an eye-catching announcement to herald the coming into existence of the GLC on 1 April 1965 — I was still in Berkeley — he immediately 'invented' the Motorway Box, the innermost of four high-capacity roads around Central London, which became Ringway One. In addition to its unfortunate name, it actually looked like the shape of a coffin and Londoners did not want it. In particular, after seeing the Westway developed, it became anathema to local residents.

Lesson — Early examples of transport and politics.

But, more importantly, as the Motorway Box was an 'idea' rather than a researched proposal, many residents were justifiably concerned about where exactly the motorway would run. A central concern, all over London, was that there would be 'blight' in any area due to have a motorway passing through, and housing prices would tumble. The lack of concrete plans fed this anxiety and led to the ultimate demise of the proposals.

The motorway plan was subject to major public inquiry examination as part of the Greater London Development Plan. Some small part of the three inner ringways emerged in 1973, but they were gradually killed off as a result of public opinion. More specifically, the reality was that the total plan could and would not ever have been able to be financed, which I fear should have been clear from the outset.

After I left, the GLC Establishment Committee (1969), in a report on my resignation from the Department of Highways and Transportation, effective 31 August 1969, described my early role and went on,

> "In July 1967, following a review of the department, Dr Ridley was appointed Chief Research officer, responsible to the Chief Engineer (Transportation) for the department's research activities. In a little under four years' service Dr Ridley can lay claim to a creditable list of achievements. On joining the Council's service Phase III of the London Transportation Study was just starting. He supervised the work of the consultants for practically the whole of the study, prepared the final report which is to be published by the Council under the title 'Movement in London', and devised techniques for the Council to take over the work and to develop it."

The report ended by referring to my insistence on obtaining from others the same high professional standards that I set for myself, bringing to the work zeal, ability and integrity. The Tyneside PTA was, it said, fortunate to have secured my services.

## Bibliography

GLC (1964), LTS: Existing traffic and travel characteristics in Greater London.

GLC (1966), LTS: Future traffic and travel characteristics in Greater London.

GLC Establishment Committee (1969), Report on the resignation of Dr Ridley.

ICE (1968), Transportation engineering, in *Proc. of the Conference Organised by ICE*, London, April 1968.

Ridley T (1967), Monorails in London, Report for the GLC.

Ridley T *et al.* (1966), The traffic demands of a National Exhibition Centre at Crystal Palace, Report for the GLC.

Ridley T *et al.* (1969), Movement in London — transport research studies and their context, GLC.

Ridley T *et al.* (1970), The London Transportation Study and Beyond, in *Regional Studies*, Vol. 4, pp. 63–71 and 81–83.

Waters B (1967), Get our cities moving, Conservative Political Centre, London.

# 4

# Tyne and Wear Metro 1969–1975

**How did a clear statement of objectives by government, together with a well-conceived Regional Transport Study, lead to a commitment to go ahead with one of the first new pieces of urban rail investment in the United Kingdom (UK) in many years?**

Some lessons

- A job application is greatly assisted by doing extensive research beforehand — and do not be modest.
- Career advancement may involve significant risk-taking.
- A clear statement of objectives is essential.
- The value of good communication with the public and other professionals cannot be overstated.
- Good teams can be as important as good plans.
- The importance of political unanimity and the 'time being ripe'.
- Make sure that the plan you are inheriting is right.
- Public participation can be helpful.

## 4.1 Job Change

An advertisement appeared in the *Sunday Times* in mid-1969 for the role of DG of the Tyneside PTE which would be created on 1 January 1970, under the political control of the Tyneside PTA.

Time with the Greater London Council had been a wonderful experience. I had worked with stimulating colleagues in the GLC and the government, with whom paths would cross over the next 40 years. But, at the age of 35, I was ready for a change.

Three other PTEs were being created at the same time — Greater Manchester, Merseyside and the West Midlands. But the North East was an area I knew from childhood, and it seemed a suitable challenge, as a possible next step.

I did my homework, studied the role of the PTAs and PTEs and caught up with the current North East scene. A good reference was necessary so I approached Ralph Bennett, who had been in Manchester and, by now was in a senior position at London Transport. Ralph was delighted to give me a reference but required something from me. "Let me have the draft of the reference you would like me to write for you," he said. "If I don't like it I will modify it before I submit it. Don't be modest. I can tone it down, but I can't tone it up." This was a practice that I subsequently encouraged all of my colleagues to follow whenever they asked for a reference. I also met with Wilfrid Burns, previously the City Planning Officer of Newcastle Council, to pick his brain and to discuss the 'movers and shakers' in the area.

Lesson — A job application is greatly assisted by doing extensive research beforehand — and do not be modest.

It was a subsequent shock to my system to discover that the job included negotiating with the Transport and General Workers Union. Working with the unions was always challenging, as they carried the option of bringing work to a standstill with strikes. It was therefore important to understand their perspective.

### 4.1.1 Interview

I have particularly vivid memories of the Interview and the process that followed. The interviews took place in the Civic Centre in Newcastle on 29 May 1969, and were carried out by the whole PTA, with Alderman Andrew (Andy) Cunningham in the chair. He was the Regional Secretary of the General and Municipal Workers Union, and had held many other positions, including Chairman of the Tyneside PTA and Chairman of the Durham County Police Authority.

There were eight candidates for the job of DG. As my name was last in the alphabet, I was the eighth. I walked into a large room and sat at the end of a long and wide table with Councillors along each side. At the far end sat

Alderman Cunningham. After welcoming me and letting me settle, he said, "Dr Ridley, you've read the terms of the appointment?" "Yes." "If we offer you the job, will you accept it?" My whole life passed before my eyes, my previous experience having been to be flown to London, via Boston, by the GLC, offered a job and invited to return to California to discuss with my wife whether I wanted to take it. I mumbled something about needing to look at some details, but essentially said yes. This was certainly a risk, and not the last time that I had to take a risk in climbing the 'career tree'.

Lesson — Career advancement may involve significant risk-taking.

A 20-minute interview followed, and I was asked to wait outside where, by now, the other seven candidates had gathered for the result. After 10 minutes, the Secretary of the PTA came out and invited me back into the room. I expected further questioning. "Congratulations," said Cunningham, "you've got the job. You're meeting the press in 15 minutes." No risk there! A Councillor came up to me and shook hands. "Congratulations, lad," he said, "at least you'll understand the language." Andy shook hands. "I've got a doctor in the family" he said. He was very proud that his son John had a PhD in Chemistry and who later became a Cabinet Minister under Jim Callaghan, and then Lord Cunningham of Felling.

**PHOTO 4.1. Alderman Andy Cunningham, a Great Support for a Young DG.**

*Source*: By kind permission of Lord Cunningham of Felling.

It transpired that Neville Trotter, one of the members of the interview panel, was keen to learn about my background and the PhD. I was puzzled by a request from the Newcastle City Librarian asking for a copy of my thesis, which I sent to him. Apparently, Trotter, who was the Chairman of the City of Newcastle Transport Committee, had asked for it, which a journalist told me he described as my job application and "all bloody algebra!" He was about to be 'dispossessed' of *his* buses, because all Municipal buses within the Tyne and Wear area were to be transferred to the PTE.

### 4.1.2 Moving North

I left the GLC at the end of August and went north for my first 'sharp end' job. For three months I had neither a home nor an office, so I moved into the Swallow Hotel, which served as both. When we bought our house in Rowlands Gill I became, if not the owner of a railway, rather the zero ground rent landlord. The leader of the team of enthusiasts that ran the Derwent Valley Railway, through our seven acres of land, was a neighbour who described it in detail in an article in *Model Engineer* (Swan, 1969).

### 4.1.3 White Paper: Public Transport and Traffic

The first move that led to the decision to build the Metro was not made on Tyneside, but within the government in London. In the mid-1960s there was a new surge of interest in urban problems in the UK, particularly transport problems, led politically by the outstanding Minister of Transport, Barbara Castle. In the words of the 1967 White Paper *Public Transport and Traffic,*

> "Britain is basically an urban country. The quality of urban life depends on the excellence of the transport services. The freedom to move easily about the city is something of great value. Yet this freedom can, if ill-used, do great damage to the quality of urban life. The nature of urban transport systems must be based on our ideas of the kind of cities we want. But city planning must be based on a realistic appraisal of what can be provided in

the way of transport investment. General planning and transport planning must be carried out hand in hand."

"All the studies carried out so far, from the Buchanan report (Buchanan *et al.*, 1963) onwards, suggest that our major towns and cities can only be made to work effectively and to provide a decent environment for living by giving a dynamic new role to public transport as well as expanding facilities for private cars."

"We must see how far public transport can offer new opportunities in the renewal of our urban areas. The pattern of the growth of London in the past century was largely determined by the building of the suburban railways and tubes; the structure of many provincial cities reflects the pattern of the electric tramways. New rapid transit systems for our major cities might provide an attractive basis for new patterns of development."

This was great encouragement for a young engineer/manager looking for a chance to 'get things done'!

PHOTO 4.2.   Barbara Castle, an Excellent Minister of Transport.

The White Paper was followed by a Transport Act (HMSO, 1968) among whose provisions was the creation, then, of four PTEs centred on Birmingham, Liverpool, Manchester and Newcastle. These came into being late in 1969 under the control of *ad hoc* bodies of politicians nominated by individual local councils, the PTAs) because, at that time, the reorganisation of provincial local government had not yet taken place.

The White Paper had a clear idea of what was required of an Executive which, in the case of Tyne and Wear (then Tyneside), consisted of a DG and Directors of Operations, Finance and Planning:

> "Its primary job will be to plan the public transport system of the Area as a whole in the context of the development and traffic plans of the local authorities. The Executive must comprise men of vision and wide experience; and they must employ staff skilled in the latest techniques of transport planning and development not only by road, but by all means of transport. It will be the job of the Executive to work out with the local authorities a practicable balance between private and public transport, to integrate the bus and rail services in the Area and to evaluate the costs and benefits of major new investment in public transport, whether in fixed track systems, reserved routes for buses or by other means."

I have always taken the view that most success is broadly based 50 percent on good management and 50 percent on good luck and timing, but I also believe successful projects are rarely, if ever, lacking in a clear statement of objectives. Would that all government policies in subsequent years had been founded on such a clear statement of objectives as those in the White Paper. So far as the Metro was concerned we were fortunate in our timing, in that there was really only a two-year window of opportunity for us to get the money for the Metro. Neither before nor afterwards was money available, as Manchester found to its cost. This was the first time that I became aware of the fundamental need for clear objectives in order to achieve project success, one of the most important lessons of my lifetime, which, happily, was repeated immediately in my next assignment in Hong Kong.

Lesson — A clear statement of objectives is essential.

## 4.2 Beginning

Work with the PTE began on 1 September 1969. We were to take over responsibility for the municipal bus services of Newcastle and South Shields and it seemed best to have open and frequent communication with the public. So on 1 January we held a photo opportunity (photo-op) in South Shields, with some of their buses. Each town had different colours — yellow (Newcastle) and blue (South Shields). It was decided that all buses would become yellow. So in South Shields, on the first day we became responsible for the service, and there were repainted buses available to be photographed. On meeting several customers, I asked one elderly lady what she thought of the newly painted buses. "Well," she said, "I'm not sure about the colour, but they're certainly warmer than the old blue ones." What seemed important to her, was that we had talked together, and she was able to express her opinion frankly. Such connections cannot be over-rated.

I had learned through good friends who were themselves journalists, that it would be helpful if the press were kept aware of what was happening and could explain what was going on. In this way the media would be well informed and so would be the local community. Because of a commitment to good communication with the public, regular appearances in the media were a necessity as plans were developing. This meant that I became well known in the area. When on TV I rapidly learned, not how to speak, but what to look out for; how to look out of the corner of my eye to watch the producer, behind me, counting down the seconds with his/her fingers to the end of the interview, thus allowing me to snatch the last word. I soon discovered how much the public took in, or not, from watching television. "Saw you on television last night," a bus conductor, hairdresser or a waitress would say. "Oh, how was I," I would enquire. "You were great." "What did I say?" A look of embarrassment, "Well, I was busy putting on the kettle at the time." Thus, I learned the value of good communication, but that being prepared to talk to the press was sometimes more important than what I actually said, in terms of public perception.

Lesson — The value of good communication with the public and other professionals cannot be overstated.

As with any transport boss, a variety of messages arrived by post. One elderly lady wrote, from Gosforth, about how bad the bus services were. After 31/2 pages of complaint she finished, 'And by the way, I'd like to know what you are a Doctor of. You're certainly not a Doctor of running buses.'

### 4.2.1 The Structure of the New PTEs

An early opportunity allowed me to explain the new structure of PTEs, as I saw them. The Department of Civil Engineering at the University had organised a three day Symposium on 'Tomorrow's Bus World,' and asked me to speak about the role of the PTE's (Ridley T, 1970). This gave me the opportunity to discuss the 1967 White Paper and the 1968 Transport Act, and the roles of the PTA and the PTE.

> "The function of the Authority is to lay down the general and financial policy to be followed by the Executive. The Authorities consist both of members nominated by constituent local authorities and, in addition, appointees of the Minister. The PTEs have three main functions. First, they must operate the bus services in their Area that were formerly municipally owned. On Tyneside there were only two such municipal undertakings, those of Newcastle and South Shields. (Sunderland was added later). The second function is to reach operating and financial agreements with the National Bus Company (NBC) and British Railways for the provision of services in the Area. The third function is to produce a Plan for the development of public passenger transport in the Area."

## 4.3 Policy Framework

### 4.3.1 PTA and PTE Joint Policy Statement

The Executive and Authority were also required, as one of their first duties under the 1968 Transport Act, to prepare a Joint Policy Statement (Ridley T *et al.*, 1970) that set out the general policies they would adopt.

The Foreword to the Statement was signed by Andy Cunningham and myself. Interestingly, it was a joint statement and not prepared by the PTE for the PTA, which would have been the normal way in local government. It said,

"Tyneside is changing. ... The task we have is not easy. The services provided at present are far from satisfactory whether by bus or by rail. The cost of providing public transport is rising rapidly. Some hard financial decisions have to be taken, that will affect everyone on Tyneside. But Tyneside must have a public transport system that is modern and efficient if it is to obtain full benefit from all of the other developments that are taking place. The prime responsibility is ours but we cannot succeed alone. We will work with other bus operators and British Rail, as well as the local authorities to provide an efficient system. This will not be achieved overnight but we are confident that there is a will on all sides to give Tyneside public transport of which it can be proud."

From the outset of occupying senior positions during my career I have believed that the chemistry of relationships is an absolute prerequisite to success, whether dealing with strategy making, project management or other subject. The authors of the Transport Act must have thought so too. The Executive (Ridley, Hurst, Fletcher, Howard) were all unknown to each other when appointed. The PTA, a brand new body, had not worked with the newly formed PTE. I can say with confidence that everything we achieved, and it was considerable, during the five years of my involvement, sprang from the relationships and understanding, and the setting of objectives, which we developed through working on the drafting of the Joint Policy Statement. It was a lesson that stayed with me for the rest of my career. This too, was another lesson that I learned again in Hong Kong.

Lesson — Good teams can be as important as good plans.

### 4.3.2 Tyne and Wear Plan

Prior to the creation of the PTEs, the Tyne and Wear area had become one of the last metropolitan areas to follow the fashion of the 1960s by carrying out a comprehensive planning and transport study. Commencing in January 1969 the consultants — the American Alan M Voorhees and Colin Buchanan & Partners — produced the 'Transport Plan for the 1980s (Voorhees *et al.*, 1971a), with a Plan Committee of local authority representatives in the area. The road proposals were essentially those that had already been developed previously and were, by and large, accepted and recommended by the consultants. The new concept, indeed the heart of proposals for improvement of public transport, was that a new underground tunnel should be built to carry a rapid transit system so that local rail services could pass directly through Newcastle and Gateshead, linking up with existing lines to Tynemouth and South Shields on the north and south banks of the Tyne. Even more significantly, perhaps, it was proposed to change the balance of transport investment giving public transport a much higher proportion than previously.

I joined the Plan Committee in 1970 after assuming my DG role, under the chairmanship of Lord Ridley, no relation to me but elder brother of future Cabinet Minister Nicholas Ridley. He also represented Northumberland CC. Matthew Ridley, fourth Viscount, was known as one of the few Tories to be welcomed regularly at the annual Durham Miners' Gala. However, he was also known as occasionally forthright and had said of the Tyne and Wear Metro that it was 'the biggest waste of public money since the building of the pyramids,' a bit rich when he had chaired the Plan Committee that first proposed it.

Brian Martin had left the GLC shortly after I did and became MD of Voorhees' UK arm, AMV Ltd., so we resumed our relationship but with different roles. The Chairman of the Technical Committee, of which David Howard and Jim Hurst became members, was the City Engineer of Newcastle. He was focussed on the need for new roads, and seemed little interested in the proposal to develop rail transport. The proposals which he had developed for roads were included within the Plan. However, when monies were finally allocated, the public transport proposals were implemented, but none of the roads.

Among other members of the Plan Committee was Tom Beagley, representing the Ministry of Transport, who became another of my mentors. Soon after my appointment I started to receive a string of telephone calls from him. I felt very flattered at my young age, not realising that his purpose was good communication. He very effectively kept his finger of the pulse of each of 'his' new PTEs in this way. However, we also became lifelong friends.

## 4.4 Making It Happen

It was the consultants who 'invented' the idea of the Metro. I have always rejected the title that friends and press tried to thrust upon me — Father of the Metro. I have, however, happily accepted that I was the mid-wife, who made it happen, the essence of what engineers are about.

What no one had realised at the time was that a combination of the realisation that the Northeast had done rather well out of the road programme in the non-urban areas earlier in the 1960s, together with the arrival of the 'motorway revolt' from San Francisco via London, would kill the motorway programme some few years later. This left the Metro with an even higher proportion of investment than had been intended.

One might call this a 'marriage made in heaven' — a political PTA and professional PTE in search of a Plan, and a Plan that needed someone to drive it along and implement it, in an atmosphere created by the central government of trying to encourage and develop public transport, backed up with money available to assist in the purpose. Not quite the 'accident view of history,' but it is certain that, while (wo)men occasionally create history, they are immeasurably helped if the time is propitious.

Lesson — The importance of political unanimity and the 'time being ripe'.

Unusually the PTA and PTE were required to produce a plan *jointly.* We called it a Plan for the People (Ridley T *et al.,* 1973), thus deliberately focussing on the customers rather than engineering hardware. There was an additional reason for its attraction. Under the terms of the 1968 Transport Act, the local areas were to assume responsibility for decisions about, and payment of subsidy for, BR's local passenger services, albeit with a sliding scale of grant from the central government. As the passengers carried represented only 4 percent of the daily public transport passengers, and the subsidy would have to come from the rates (property taxes), this represented a no-win situation for Tyneside.

It seemed unwise to pay an increasing subsidy for a poor service on lines that had actually been downgraded by BR (having been electrified they were converted to diesel operation during the 1960s for reasons of lack of capital), and would also have been politically controversial. To have taken the opposite course of telling the government and BR that the lines should be closed would have caused an uproar. So the third alternative of development was pursued, as the best available course of action.

Three other major steps had to be taken. The PTE had itself to be convinced of the wisdom and feasibility of the Metro project. As a concept it looked attractive, but that was clearly not enough. Secondly, a government infrastructure grant had to be obtained and, thirdly, statutory powers to

**PHOTO 4.3.** Plan for the people.

allow the scheme to proceed had to be obtained by means of a Private Bill in Parliament.

Working together with representatives of BR and the government, the PTE sponsored the North Tyne Loop Study (Voorhees *et al.*, 1971b). It was important that we only proceeded with the Metro, the concept for which had been developed by Tyne and Wear Plan and which we had joined late in the day, if we had fully convinced ourselves that it was, indeed, the best of several possible options. The Study made a comparison of upgraded rail, a bus-way replacing the railway, a conventional all-bus system and a light rapid transit system. Particularly relevant to the ultimate decision to endorse the latter, the light rapid transit development, was the fact that under the 1968 Transport Act, the government had for the first time made financial support available at the same 75 percent grant rate as for highways. The report addressed our BR dilemma and said that,

> "The government is paying grant to British Rail for a number of un-remunerative services in the area of the PTE. More than half of the present grant aid in the area is to support rail service on the North Tyne Loop (NTL). … Early in 1971 the PTE must decide whether or not to accept financial responsibility for this service. The purpose of this study has been to determine whether the most satisfactory transport service for the area served by the NTL can best be provided by retaining the existing system and continuing to pay grant or whether some alternative transport system would be better."

Thus, we convinced ourselves.
Lesson — Make sure the plan you inherit is the right plan.

In the event we decided that rapid transit was the best option. This meant that an application for such a grant had to be made to the government — fast progress (Ridley T to T Beagley, 2 July 1971). The approach was at three levels. First, there was close cooperation between transport planners and economists within the PTE and their opposite numbers in the government so as to ensure that the presentation matched the style required for assessment. Secondly, close contact was maintained between the PTE Directors and the senior civil servants who received the recommendations of their technical colleagues and then had the responsibility of deciding whether to recommend the project to the Minister. Finally, a united political stance was taken by Councillors and MPs of both the majority Labour and minority Conservative parties.

### 4.4.1 British Rail Uncertainty

On 11 October 1972, we learned that our application had received an agree-ment in principle to the payment of 75 percent of the cost of preliminary expenditure. On 3 December, David Howard informed the Department that I had been told that the people in BR, with whom we were working would put a recommendation to their Board that 'will endorse the rapid transit proposal in principle, subject to certain provisos about technical and financial feasibility'. However, David had warned that the message had not yet been put in writing and went on, 'I believe there is still a feeling within BR that a BR electrification scheme might be appropriate to Tyneside, if, for any reason the rapid transit proposal cannot be implemented. However, they have made one very helpful statement that, in their view, there is no alternative to the tunnel-based Tyne and Wear Plan system, which fulfils in detail the objectives of the strategy. The grant application itself (Tyneside PTE, 1972), submitted in May 1972 was able to say 'British Rail have agreed to support our Bill in Parliament and to cooperate in its preparation. Considerable effort, both at Eastern Region headquarters and locally in the Newcastle Division, is helping to resolve the technical and operational problems of introducing rapid transit'.

### 4.4.2  Securing Commitment

Less than two years after it had been first conceived, the Metro received the blessing of the then Minister of Transport, John Peyton, in the form of approval of the grant application. Undoubtedly, two factors helped the application: first, the need to take decisive action either by closing or developing the existing rail system rather than simply let it carry on in its existing, unsatisfactory fashion. Equally important was the fact that the PTE had gone a long way in discussion with British industry towards working out proposals for the development of prototype equipment. Indeed, at this time the PTE also sought, and the government agreed, a research and development grant that allowed the conversion and equipping of an existing unused railway line for use as a test track for the prototype rolling stock built by Metro-Cammell of Birmingham, the manufacturer of London's underground stock. Here, the urge to seek ways of assisting British industry so that they could turn their sights to export orders, undoubtedly played a role in government thinking.

Much lobbying and persuading was necessary to achieve the go-ahead for the Metro. The PTA had been pressing John Peyton to visit the North East, for several months, not least because he had already visited the other three PTA's. We also wanted to press forward our case for his approval of our Metro grant proposal. Replying to a letter from the Chairman, who repeated an earlier invitation and saying how helpful it would be if he could indicate publicly the government's support for the proposal. Peyton said that he certainly hoped that the project would go forward, but it was still a 'bit on the early side' to go further than his Department had already done. Furthermore, his diary was rather full up to April, but he would see what he could arrange. In the event it was arranged that he would come on 19 September. His civil servants were somewhat nervous but he, a rather laid back Somerset country gentleman, did not show it himself. The problem was that gossip and press speculation about Alderman Cunningham, which ultimately led to his going to court and to his political demise, was already under way. I was briefed by Peyton's team to stick close to him so that it would be me, and not the Chairman, who appeared with him in any photographs.

After a successful visit he left us without giving any commitment, though he was generally positive. At his request, we had taken him on the North Tyne Loop in part to reduce further any photo-op with the Chairman. Without warning he stepped off at one station to explore, and we met a member of staff in a waiting room. Peyton was peering quizzically at its peeling paint and unkempt appearance when the man, observing Peyton, said, 'yes Minister, nothing's changed around here for more than 50 years'. I could have kissed him. That was the moment I believe when, in his head, Peyton gave the Metro the go-ahead. We had talked to all and sundry in 'selling' the project, but the participation of one particular member of the public may have been crucial.

Lesson — Public participation can be helpful.

By 18 December 1972 we received the news that our grant had been approved, leading to much self-congratulation, and to hurrahs in the press, hailing 'Tyne's £65m ticket to ride,' and 'how Dr Ridley brought new trains to Newcastle'. An unhappy sequel was the news that Manchester was fighting for its £70m tube, which had been ahead of us in the race to develop a project when we set off at the beginning of 1970, but had been overtaken. We had scooped the pool, the money had run out, and John Peyton had refused the grant application for their scheme.

I wrote to Peyton saying that, in addition to the PTA Chairman having written to him about the government grant, I wanted to write on behalf of myself and my PTE colleagues to say how grateful we were for government support and 'with such speed'. I had always believed that if the Northeast could initiate well-conceived proposals then it could anticipate government backing. This view had been wonderfully vindicated. His reply charmingly said that I could take it that the government's decision was in no small way an expression of confidence in myself.

Meanwhile it transpired that BR had been working, separately, on an alternative to the Metro. As luck would have it, bad luck for them, it landed on the desk of the BR Chairman, Bob Reid, just as he was able to read press reports of Peyton having blessed the PTE's Metro. BR were professional friends and collaborators, and they did help us with implementation, but I always felt that they did not take us seriously, and regarded us as schoolboys who were incapable of getting the go-ahead, let alone completing the project.

About this time, Tom Beagley, was promoted to a Deputy Secretary position in the Department of Environment. In response to my thanking him for all the help and guidance he had given me, he said how much he had enjoyed working with us over two years. Nobody was more surprised than he when I was appointed as DG, but 'the very fact that I now think of you as one of the leading men in the public transport field shows what a splendid success you have made of the job. We are all terribly happy about the way things have gone on Tyneside and this, I am sure, is due not only to you but the very good team you have collected together.' (T Beagley to T Ridley, 7 January 1972).

A campaign was needed to promote the project and Tom Bergman, a larger than life Czech who ran a PR company in Newcastle, was appointed. He had big ideas and soon rejected my proposed leafletting campaign. "We will run a campaign to get both public and political support," he said, which is what we did. We spent much time in London talking to MPs and developed a united front. Labour and Tory, were normally at each other's throats, but at one in wanting to get money for the Northeast out of the government. It worked, both locally and nationally.

Additional money was needed for a test-track in Longbenton to the north of the Tyne and a special application was made to the Department of

Transport. We chose the Longbenton site after trying to persuade residents living adjacent to a disused railway line in South Sunderland to allow us to convert it into a test-track. In this way Sunderland would have a small part in the Metro action. After a rowdy and negative public consultation meeting we 'changed the plan'. Subsequently the Longbenton test-track was, fortuitously, an important part of the UK's proposal for a package of E&M contracts for the Hong Kong Mass Transit Railway.

### 4.4.3 Parliamentary Bill

Now the Executive took the next step of obtaining Parliamentary powers for the construction of the Metro and the Tyneside Metropolitan Railway Bill was deposited in November 1972, shortly before government grant approval was given. The Bill received Royal Assent in July 1973. There were a number of petitions against the Bill, as is usual, from both private and public bodies, but all were satisfactorily settled. My Proof of Evidence (Ridley T, 1973a) is included in the Bibliography.

I was examined before the Bill Committee on 1 May 1973, during which I was questioned about the project. I recall that it was a gentle canter, but good practice for the more torrid experience at the later King's Cross Inquiry. Among other things I quoted (Ridley T, 1973b) the important words of the consultants to the Tyne and Wear Plan Committee,

> "We believe the Tyne and Wear Plan area is in a position to avoid some of the difficulties encountered in other urban areas in their transport planning policies. At present car ownership is lower than in many other areas. There are clear signs of the drift away from public transport, but our analysis shows that the area cannot afford to let this continue as no reasonable level of highway expenditure could meet the resulting demand. The opportunity should be taken, therefore, to maintain and improve the public transport system while ridership is still relatively high. While the bus system will always be an important element of the public transport services we believe that the only way to provide the level of service and accessibility required in the future, in the face of increasing road traffic, is by significant investment in a system with its own exclusive right-of-way. This system should be based on the current railway facilities with appropriate renewal and upgrading of lines and stations and the

improvement of central area accessibility by linking of rail lines north and south of the Tyne."

Thereafter, Mott, Hay and Anderson began detailed design and with the first contract being let in October 1974, shortly before I left for Hong Kong, tunnelling works commenced less than four years after the first thoughts on the scheme had been developed by the consultants who produced the Tyne and Wear Plan. By any standards this represented very rapid progress.

The following chapter explains how I came to leave Tyneside and start another metro in Hong Kong. In a farewell piece a young reporter Quentin Peel, later an eminent *Financial Times* journalist, wrote in the local press under the heading — 'Mr Transport is off to the Far East' (*Newcastle Journal*, 1975).

> "When Tony Ridley came to Newcastle to take over a reorganised bus system, he was a backroom boy. Five years ago, at the age of 36, he was described as a brilliant young boffin, coming as the first head of combined bus services in Newcastle and South Shields. Now he is to leave Newcastle with the groundwork laid for a new system which will link everyone from Whitley Bay to Sunderland and Blaydon to South Shields. He is staying long enough to see work started on the Metro system for Tyneside, before he flies off to Hong Kong to lay the groundwork for another whole organisation of public transport."

## 4.5 What Happened Next?

Geoffrey Skelsey (2000) wrote a three-part review of the Metro in Tramways and Urban Transit, the journal of the LRTA. He opened Part 3: Deregulation and Expansion as follows:

'One of the saddest aspects of the T&W Metro's history is the brevity of the period during which it was able to establish its real potential. The core network was completed in March 1984. Little more than two years later, on 26 October 1986, the Transport Act 1985 came into effect and passenger road transport services were deregulated. At a stroke, the central plank of the Tyne and Wear's transport strategy was removed'. Amen to that.

The LRTA, formerly the LRT League, is an organisation of 'enthusiasts'. They approached me in 1973 to become their President, to which I readily agreed, not least because they had been 'advocating better public transport since 1937'. The responsibilities were not onerous, principally to attend the AGM and deliver a Presidential Address which, surprisingly, I managed to do very regularly except during my time in Hong Kong, until I resigned in 1992 after 19 years. In the first address I said, *inter alia*,

> "Every transport problem is different and demands a different solution, or a combination of solutions — the light railway, the tram or the Supertram as some people have called our system; or the Supertram combined and integrated properly with other modes. I see from the Times of today that someone was pressing the case of the trolleybus. None of these are a panacea to the urban problems we face. What I think the League can legitimately do, and I certainly urge it to do, is to demand a proper recognition of the possibilities that are presented by light railway technology. In different circumstances we could easily come to different decisions about the Northeast of England. In the event we came to a decision and pushed it through, and got support from local and national politicians for our particular network." (LRTA, 1974)

## 4.6 The Millennium Bridge

After leaving Tyneside for Hong Kong, I never went back to live there, but did return from time to time. I was invited to the Royal Opening of the Metro in November 1981. In 2002, I was overseas and unable to accept an invitation to the opening, by the Queen, of the extension to Sunderland.

A letter from the Director of Engineering Services of Gateshead Metropolitan Borough Council in September 1996, shortly before the end of my ICE Presidency, said, 'You may be aware that a design competition for a new footbridge across the River Tyne between the eastern quaysides of Newcastle and Gateshead is being promoted .... The Council would therefore like to invite you to take the chairmanship of the panel which will determine the winner of the competition'. I immediately said yes.

We held our Panel meeting in Gateshead in February with six shortlisted designs. In the event the decision was that the Gifford/Wilkinson 'winking eye' was outstanding. The only issue was — would it work? Upon satisfying ourselves on this point, we announced our decision, to general acclaim.

Thus, it is a matter of some pleasure to have played a small part in adding three bridges to Tyneside's historic collection — the Queen Elizabeth II Bridge over the Tyne, and Byker Bridge for the metro and the Millennium pedestrian bridge.

**PHOTO 4.4a.   Queen Elizabeth II Bridge.**

*Source*: By kind permission of graemepeacock.photographicimagery.

**PHOTO 4.4b.   Byker Bridge.**

*Source*: By kind permission of graemepeacock.photographicimagery.

**PHOTO 4.4c.   The Millennium Bridge.**

*Source*: By kind permission of graemepeacock.photographicimagery.

# Bibliography

Buchanan C D *et al.* (1963), Traffic in towns. A study of the long term problems of traffic in urban areas, HMSO.

LRTA (1974–1990), Ridley Presidential Addresses, in *Modern Tramway and Light Railway*, Ian Allan Ltd, Weybridge.

Ridley T (1970), The role of the passenger transport executives in 'tomorrow's bus world', in *Proc. Symposium on Tomorrow's Bus World*, pp. 51–58, University of Newcastle.

Ridley T (1973a), *Proof of Evidence*. Tyneside Metropolitan Railway Bill.

Ridley T (1973b), *Examination*. Tyneside Metropolitan Railway Bill, pp. 22–38.

Ridley T *et al.* (1970), Joint Policy Statement of the Tyneside PTA and PTE.

Ridley T *et al.* (1973), Public transport on Tyneside — a plan for the people, Tyneside PTA.

Ridley T to T Beagley (1971), Application for Infrastructure Grant, 2 July.

Skelsey G (2000), Twenty years of the Tyne and Wear Metro, in *Tramways and Urban Transit*, p. 296, Light Rail Transit Association.

Swan K (1969), The Derwent Valley Railway, in *Model Engineer*, Vol. 135, No. 3369.
Transport Act 1968, HMSO.
Tyneside PTE (1972), Tyneside Rail Rapid Transit, Application to the Department of the Environment for an infrastructure grant.
Voorhees A M *et al.* (1971a), Transport Plan for the 1980s.
Voorhees A M *et al.* (1971b), *North Tyne Loop Study*, Vols. 1 & 2.


## Press

*Newcastle Journal* (1975), 'Mr Transport is off to the Far East', Quentin Peel article.


## Personal Correspondence

Beagley T to T Ridley (1972), 7 January.

# 5

# Hong Kong 1975–1980

How did the Hong Kong government go about its planning and decision-making for the creation of the Mass Transit Railway Provisional Authority (MTRPA)? How did the MTR Chairman save them from a near disastrous decision to sign a 'letter of intent' with a contractor consortium? Why was the MTR Modified Initial System a success? Who proposed the policy of building on top of the railway?

Some lessons

- Always be prepared to talk about offers of employment.
- Talk to mentors, if possible, when faced with major changes.
- Good interpersonal relations between client and contractor are crucial.
- 'On time and within budget' is highly dependent on when they are calculated.
- Every project needs a champion.
- Urban rail projects represent the management of complexity.
- Single solutions to transport network problems are likely to be counter-productive.
- Attention to good process is central.
- Accepting the lowest price bid for a single contract can be counter-productive.
- Revenue forecasts are more likely to be erroneous than time and cost.
- An intelligent and efficient client is very important.
- The task of the MTR project was NOT to deliver infrastructure, but to create a successfully operating transport service.
- A mid-career break can be very refreshing.

## 5.1 A Surprise Approach

### 5.1.1 Recruitment

On 15 May 1974, the phone rang in my office at the PTE in Newcastle. It was a head-hunter, Michael Egan of PA Management Consultants who says "Dr Ridley, you will know of Hong Kong?" "Of course." "You have heard of the Hong Kong MTR?" "Yes." "Would you be interested in working for the MTR?" "No." "Why ever not?" To cut a long conversation short I agreed to visit Hong Kong simply to find out more about it.

I had not committed myself to moving there but had agreed to be interviewed. It can never be wrong to talk about a job opportunity. It is easy to say no, if an opportunity turns out not to suit your liking. Lesson — Always be prepared to talk about offers of employment.

When I joined the Tyneside PTE in 1969, the Metro had not been thought of. The job did not include such a project. But, by virtue of having championed it, sought Parliamentary powers and finance, and let the first contracts, it had become what the press wrongly called 'Dr Ridley's baby'. Hong Kong was altogether different. The HK government had already done a great deal of work planning the project, with the help of British consultants, before I arrived.

I flew to Hong Kong on 12 July, not exactly enthusiastic about the possibility of leaving Tyne and Wear. Five of us were shortlisted for the job of MD. The government was already close to appointing the Chairman-designate, Norman Thompson. Alan Cotton was being lined up for Operations Director-designate, and the government had Lau Wah Sum in mind for the finance job.

The Financial-Secretary (FS), chaired the appointing board as chairman of the MTR Provisional Authority, together with several government colleagues. Anthony Bull, retired Vice-Chairman of London Transport and advisor to the British consultants — Freeman Fox and Partners, together with Kennedy and Donkin — was also advising the FS on appointments. The interview seemed to go well.

I was given a thorough briefing on the project and, very importantly, was taken out to dinner by Anthony Bull (a mentor) at the Repulse Bay Hotel. He quickly ascertained that I was very interested, but had qualms about walking away from the PTE and the Metro. He made it clear that, in his opinion, Hong Kong needed me more than Newcastle T&W did. He also had a major interest in ensuring that, given that the Chairman (designate) was an accountant, the MD should be a transport man. Well, if the great man (and mentor) thought that I could leave Newcastle with a clear conscience, who was I to argue?

Lesson — Talk to mentors, if possible, when faced with major changes.

I was offered the job as MD but also given the opportunity to return with my wife before signing up. Jane and I went on 21 October, luxuriated in the Mandarin Hotel, were shown schooling and housing, fed dim sum and otherwise entertained, and I signed up. The contract included a side letter that would reimburse me for one year's salary should the project fall through.

I had not accepted the position without a great deal of heart-searching. The work in Tyne and Wear was not finished, but we had done a lot in the previous five years. Construction of the rapid transit system had started and the region would have a first-class, speedy and reliable service of public transport in the foreseeable future. I knew that my colleagues would successfully complete what we started together. Indeed they did and Hurst, Howard and Fletcher succeeded me in turn as DG. I joined the MTRPA when I arrived in Hong Kong in February 1975.

Before Norman Thomson left London for Hong Kong as the Chairman-designate, and I had been appointed as the MD, we met at the RAC Club for dinner. We talked about everything to do with the project. One thing he insisted was particularly important, "when we are letting contracts, you and I are going to interview the Chairman or the MD of every shortlisted bidder and ensure they give their personal commitment to deliver the contract successfully." We carried this through. There were a number of occasions when we had to remind them of their promises. This was one of the important lessons I learned — it is too late to call in the Senior Manager of a contractor when things have gone wrong, and there is already a 'dispute'. Get to know them personally, at the outset when and before they have won the bid, and they are eager to cooperate. The value of good relations between client and contractor runs from the beginning to the end of a contract.

Lesson — Good interpersonal relations between client and contractor are crucial.

I left the PTE and proposed to fly out to Hong Kong at the beginning of February, with the family to follow at the end of the school term. Thompson wisely advised that the timing would not be particularly good, because of Chinese New Year, and suggested that I spend time visiting London Underground and also helping the newly appointed Director of Engineering, Paul Andrew, who was busy recruiting staff. Paul and I were shown around LU by Charles Cope, the Operations Director, whose boss I subsequently became five years later.

## 5.2  Genesis of the Mass Transport Railway

The Hong Kong MTR has a high reputation. Some call it the best in the world, though several more recent systems may claim to match it. It certainly introduced a new higher level of project management but, most particularly, it had a uniquely effective *client*.

A crucial part of this project is the way its objectives, planning and financing were developed. I have told the story of the early project several times — Ridley T (1976, 1977a, 1978, 1980a), which I shall not rehearse here. Mike Winney, Editor of NCE, gave it full coverage in Construction News (1978) subsequently. Staff members and consultants published papers in *HK Engineer*, The Journal of HK Institution of Engineers. ICE published a suite of papers, led by FF&P (Edwards *et al.*, 1980).

### 5.2.1  The Mass Transport Study

In the mid-1960s Hong Kong, although far from having overcome the housing problem created by the massive influx of immigrants from China over a 15-year period, began to examine its transport problem. Following survey work by the (then) Road Research Laboratory from the UK, a Mass Transport Study was commissioned with the object of developing, 'the best solution to Hong Kong's long-term mass transport problems, consistent with planning goals, development plans and a level of mobility that will allow the Colony to continue to prosper.'

The Study, by Freeman Fox and Wilbur Smith contained,

"recommendations for a long range mass transport programme, the principal item of which is a 40-mile rail rapid transit system. We have devoted a major portion of the report to this underground and overhead rail system. Our proposals for this system take the form of a six-stage development plan aimed at opening the first section of line in 1974 and the last in 1984. To achieve this, detailed planning and design must start in 1968 and construction in 1970. The scheme has been worked out so as to achieve the shortest completion period consistent with the economy."

The early estimated total cost of the system (at 1967 prices) was HK $3,400 million. While very substantial revenues were forecast, 'some assistance from public funds in the early years is likely to be necessary'. It is noteworthy that the report also said 'the surface and rapid-transit systems should be complementary rather than competitive, each serving the type of travel it is best suited to accommodate, and producing together an effective and efficient transport system.' That certainly fitted well with the so-called 'best integrated urban public transport system in the UK' on Tyneside.

### 5.2.2 Other Studies

The MTS was followed by a Long-Term Road Study. Thus, public transport had been put forward in advance of road proposals although, of course, a proper road system was essential for the important bus element. The MTS proposals were quickly followed by a detailed feasibility study which was reported in HK Mass Transit Further Studies (FF&P, 1970).

A full Preferred System of 52.6 km was recommended, costing some HK $4,400 million (at 1970 prices), to be developed in nine stages. This cost did not include fees, financing costs or government administration costs. An Initial System of the first four stages, 20 km long, would cost some HK $1,900 million (at 1970 prices). 'If government gives instructions to proceed with the detailed design and preparation of contract documents by mid-1971, construction could be started at the beginning of 1973 and the first stage opened early in 1976. The Initial System could be completed by the end of 1978 and the full preferred System by 1986.

It is important to note the gradual increase in cost during the planning period, but Hong Kong did not commit itself to figures until the moment when the first contracts were let. Witness also the Olympic Development Authority's undoubted success in meeting its targets for London 2012. Far too often sponsors become 'committed' to time and cost budgets too early in the project planning process.

Lesson — 'On time and within budget' are highly dependent on when they are calculated.

A period of two years passed, during which government considered the proposals. Then, in May 1972, it decided that construction of the MTR should proceed, subject to satisfactory arrangements being made for financing and for letting contracts.

## 5.3 Every Major Project Needs a Champion

Philip Haddon-Cave, was the new FS. He was an Australian who had come to Hong Kong from the Colonial Service in Africa and took considerable personal interest in the project. Indeed, one of the unique aspects of the MTR is that it depended heavily on initiatives taken by him as the HK equivalent of the Chancellor of the Exchequer.

Every major project needs a champion who, working with others, takes responsibility for driving it forward until a well-founded decision to proceed is achieved. I believe that I performed this role for the Metro in Newcastle. Haddon Cave was undoubtedly the champion for the MTR and, in my experience, also the best project client. This is particularly necessary because, as my career experience taught me, urban rail projects represent the management of complexity.

Lesson — Urban rail projects represent the management of complexity.

On June 1972, the FS said to the Legislative Council,

"The government has decided that, if the demand for movement in Hong Kong in the late 1970s and thereafter is to be satisfied, the surface public transport system must be augmented, if at all possible, by an underground mass transit railway.' But he also emphasised that the 'MTR cannot, on its own, constitute a solution to our emerging transport problem."

Happily, there were no single solution merchants in Hong Kong. This is not true everywhere in the world. Sadly, where this has not been understood, in the absence of a sound strategy – the project has become the strategy, often with disastrous consequences.

Lesson — Single solutions to transport network problems are likely to be counter-productive.

**PHOTO 5.1.   Sir Philip Haddon-Cave, a Splendid Champion and Client.**

The FS went on to emphasise that there were interconnected issues,

"first, that a substantial sum of public money can be committed to assist on financing the project; secondly, that the railway can be constructed without undue disruption; thirdly, that the system can be operated in such a way as to generate a sufficient cash flow to service the investment and cover operating costs. So the way is now clear to see whether an affirmative answer can be given to the fourth question, namely, whether sufficient outside capital finance can be obtained on appropriate terms as regards interest rates and repayment arrangements."

He was wise enough to link the financing problem with the form of tendering and the letting of contracts, and the most appropriate operational arrangements. These issues would have to be looked at separately but, eventually, brought together in a coherent scheme for the financing, construction and operation of the system.

He then set up a small Government Steering Group, under his chairmanship, to study the viability of the project. It went fully into the problems of financing, constructing and operating the MTR, including consultations with financial interests and groups of potential construction companies and equipment manufacturers. He was able to ensure that the Hong Kong Bank were willing to assist in exploring the various ways and means of raising non-government finance. A senior official of the Bank was appointed to serve on the Group.

Finally, he told the Steering Group (SG) that

"work will be starting on drafting legislation to provide for the compulsory acquisition of property (where it is unavoidable), the alteration of streets and the compensation of those people whose property will be interfered within the course of construction. The preparation of this and other enabling legislation will be supervised by the SG, which will also keep a watching brief on the consultants' activities and the progress of ground investigations and various associated works."

Thus, the tone of the development was set.

Good planning and decision-making are essential. This paid dividends in Hong Kong, just as it had in T&W. Would that all major projects could be addressed with such clarity? Furthermore, commitment to financial viability is necessary. This does not mean that 'every project must pay for itself. Part-financing by government is perfectly sensible if the money required is accurately calculated and it can be treated as income by the project client.

One earlier consideration was noticeably missing, however. There was no reference to the surface bus/public light bus and rapid-transit system being complementary rather than competitive, not perhaps because this was considered unimportant, but the concept of a total transport system, as we had developed in T&W, had become lost in the consideration of the financial aspects of the project.

## 5.4 The Japanese Letter of Intent

During the next two years international consortia put forward competing proposals for the construction of the Initial System under a single contract. This culminated in the signing of a Letter of Intent with a Japanese consortium in

February 1974, which fell through only 10 months later, leaving a vacuum into which the newly appointed MTR Chairman-designate was pitched.

The attraction of the proposal, which was not matched by the international competition, had been that the Initial System would be built for a fixed price. This was clearly a seductive proposition where the decision-making process placed financial control ahead of almost all other considerations.

The need for financial control had been illustrated during those two years during which the estimated cost of the Initial System increased two-fold and, by September, had reached (at 1974 prices) three times the Further Studies estimate of HK $1,900 million (at 1970 prices). Nor did this take any account of future escalation. To keep pace with the cost increase, the assumed fare range for the Initial System had been pushed up from HK $0.20–1.00 (at 1970 prices) to HK $0.80–2.40 (at 1974 prices).

Haddon-Cave now set up a Provisional Authority (the forerunner of the MTR Corporation), with its first meeting on a Saturday in March 1974 (Minutes, MTRPA, 1974a). It continued meeting at weekly intervals until September 1975, when the MTRC came into being. Initially, the membership was comprised of government officials with consultants and financial advisors in attendance. Directors (designate) of the MTRC joined as and when they were appointed. The Chairman-designate paid his first visit to Hong Kong in July 1974.

An early action of the Authority was to convene a Financial Management Committee, (Minutes, FMC, 13 June 1974) to carry the business aspects of the project forward. Its membership consisted of the FS; Mike Sandberg, Deputy Chairman of HSBC (later a MTRC board member and then Chairman of the Bank): Mike Barnett of Kleinwort Benson, and Launder of Wardleys, both Financial Advisors; Lau Wah Sum from government; and KY Yeung, also from government and one of Philip's high-flyers, acted as Secretary. This small committee was particularly important, given that the policy was not to proceed unless sufficient non-government funds could be raised to finance the Initial System.

## 5.4.1 A Setback: The Letter of Intent Fails

On 14 January 1975, a memorandum had been put to the Executive Council, 'Arrangements for the Construction and Equipment of the MTR'. It said,

"On 11 December 1973 honourable members advised and the Governor ordered that the Japanese Consortium's pre-emptive bid for the MTR contract be accepted. Following this the Mass Transit Steering Group undertook immediate discussions with the Consortium with a view to the issue of a Letter of Intent to enter into negotiations for a formal contract. The Letter of Intent was signed on 15 February 1974. At a meeting on 17 December 1974, the Consortium notified the MTRPA that it was unable to proceed with further negotiations for a contract to build, equip and finance the first four stages of the MTR within the framework of the Letter of Intent."

Since the Letter of Intent had been signed, 'the continuing impact of world inflationary forces' had led to the Consortium giving progressively stronger indications that it had been having difficulties with the project. The Consortium's bid had been accepted because it offered a price within the government's ceiling of HK $5 billion, without cost escalation, and held out the opportunity of an early start and completion of the project. Neither of these advantages had materialised because the Consortium had asked for a revised price ceiling of HK $6 billion, together with provision for price escalation and had put forward design proposals which were 'incompletely drawn up and were totally unacceptable'.

"By the terms of the Letter of Intent the Consortium were required to produce detailed design solutions that would meet the MTRPA's design and performance requirements. The Consortium failed to do this, but offered instead a 'Basic Plan' which would embody all detailed design solutions. The Basic Plan was not received until mid-November 1974. It was thoroughly examined by the PWD, the consulting engineers, the engineering officers (designate) of the future Corporation and the PA itself and it became clear that it was unacceptable because it was either inadequate or incomplete compared with MTRPA's specifications for engineering requirements, or the attainment of revenue objectives, or unsuited and, in some cases, almost irrelevant to Hong Kong's conditions."

The paper then described proposals for a Modified Initial System that had been developed, and the method of implementation. The Executive Council was invited to note the history of negotiations with the Consortium; to note the arguments for continuing with the railway; and to advise whether the MTRPA should proceed with the MIS on the basis of a multi-contract approach. The Executive Council agreed to the project proceeding, but with two substantial qualifications in that they were to be consulted again when initial representative tender prices were known, as well as arrangements for financing the system. Happily, the first tenders confirmed the estimates and the MTRPA were able to put together a package of finance — export credits and stand-by loans — which were enough to cover the cost of the MIS. With commendable speed the project was resurrected when it had been on the point of collapse.

Norman Thompson was instrumental in developing the Modified Initial System. This was a revised proposal for a smaller, 16 km, project. It was to be carried forward on a multi-contract basis and was estimated to cost HK $5,850 million (at current prices) inclusive of design, supervision, land and compensation costs together with the cost of inflation (estimated at 7 percent p.a.) and finance charges until construction was complete. Cash flow calculations were made, based on a fare structure of $1.00 with $0.50 increments, which suggested that the MIS would have paid back its loans by 1992 and that operating cash flow would be positive by 1983. This was on the basis of a government equity injection of HK $800 million, the remainder of the money having to be borrowed.

I have written above of my admiration for Philip Haddon-Cave. His one mistake was to have accepted the Letter of Intent. It is safe to say that the newcomers to Hong Kong, who became Directors of the MTRC, and Norman in particular, saved Philip from his fairly disastrous decision. He was fond of saying that there was nothing wrong with bureaucrats — good bureaucrats. Philip was a very good bureaucrat. His papers were always immaculate and his budget speeches, which lasted much longer than any ever given by a UK Chancellor-of-the-Exchequer, were the epitome of clarity.

## 5.5 Taking Up the Reins

Paul Andrew and I flew out to HK together, on an excellent first-class service, enhanced no doubt by the fact that I was sitting next to a senior British Airways pilot. Ironically, the movie which was showing was 'the Taking of Pelham 123', about a metro hijack in Manhattan. I was collected and taken to the Excelsior Hotel, not the Mandarin now that my recruitment was completed. I started to unpack, idly watching a local English-language TV channel, when who should appear but myself. Before I left Tyneside I had recorded a farewell interview for the local BBC channel, about my achievements and the future of the Metro project. Although a coincidence, it felt like a welcome.

At the same time it was also important to get to know David Silcock who, on behalf of FFWS&A, had produced a 6-week revenue review study at Thompson's request, to take account of the move from the Initial System to the MIS (FF&A, 1975). Thus, one of my roles became to be the 'go-to' person within MTRC on all revenue issues, themselves central to all fares questions. Meanwhile, I had attended my first Financial Management Committee.

### 5.5.1  Canton Trade Fair

An early introduction to the new world of the East arose when I and my two senior colleagues — Paul Andrew and Alan Cotton — were invited to attend the annual Canton Trade Fair in early May. We were invited to discuss what items we might buy from China, but I suspect that our hosts were very keen to inspect these 'new boys' who had come to Hong Kong to lead the project.

I reported to the 10 May MTRPA that, 'there were distinct possibilities of making arrangements for the supply of items such as cement and reinforcing bars. In addition, further consideration would be given to direct purchases of wall finishes and rails and some workshop equipment might be suitable. The provision of rolling stock and escalators seemed less likely. So far as escalators were concerned, it was necessary to ensure that the eventual suppliers undertook the installation rather than an agent. This was likely to be a difficulty for the Chinese.' A far cry from Chinese capability only 35 years later!

### 5.5.2  A Crucial Executive Council Meeting: Approving the MIS

Norman Thompson and I attended the crucial Executive Council meeting in September 1975 when the memorandum, 'Modified Initial System of the MTR' (Memo to Executive Council, 9 September 1975), was presented and approved. My admiration for Haddon-Cave had been enhanced by observing him in his office, assisted by two teams of advisors, working on two separate draft documents at the same time, the first being his budget speech and the second the above memorandum.

I presented a briefing paper, backed by the Executive Council document. The paper recapitulated the background to the project and listed the history of a series of issues considered between July 1971 and February 1975. It then described the actions taken since the approval in principle given in January 1975 and summarised the then position — cost, finance (including export credits, open market finance, and guarantees), revenue and contingences.

I have laboured some of the procedural issues, to emphasise the competence with which they were handled, in contrast to a number of other metro projects around the world that faltered, or were severely delayed, because of lack of attention to good process and decision-making.
Lesson — Attention to good process is central.

### 5.5.3 Property Development

Thompson's first major contribution to the MTR was to 'rescue' Philip from the Japanese Bid and to develop the Modified Initial System. His second was to introduce the contingency of property development, which no one in the very effective Hong Kong government had addressed. Accordingly the Executive Council agreed that the future Corporation be granted the right to engage in property development over certain stations and other sites (MTRPA (75) 24, 1974b). Approval was given to the grant of sites at Chater, Admiralty and Argyle stations (MTRPA (75) 54, 1975a), as well as over Kowloon Bay depot (MTRPA (75) 71, 1975b). The financial benefit to MTRC was important, though not massive, but increased substantially over the years when the principle was extended to new lines, particularly the line to the airport.

The competition for the first property deals was between Jardines, for years during the colonial period the most powerful corporation in Hong Kong, and Li Ka Shing. It seemed to us that Jardines were waiting for Thompson to go and entreat them to do a deal with us, while Ka Shing was a very purposeful bidder. Our practice in all contract-letting in Board meetings was for me to present the preferred and second bids for decision under Thompson's chairmanship. When Li Ka Shing was proposed as the preferred bidder for the first two developments I well remember Board member Sir Y

K Kan saying, 'we can't let the contracts go to Li; he sells plastic flowers'. So he did but, today, he is now one of the richest men in the world!

The development over the Kowloon Bay depot provided apartments for 18,000 people, and created a massive demand. One visitor to Hong Kong said to me, "do you realise Tony, that you are building Bridlington on top of the depot!" An additional benefit was that the policy produced housing for many people whose journey to work could take place exclusively by rail. One further advantage that the government had in mind was that the property owned by the MTRC would be shown as an asset in its balance sheet, which would make it easier to finance later extensions to the system.

Thus, from potential disaster the 'show' was back on the road, which led to a successful opening in 1979/1980. The project was extremely well managed, working to a very competent client. Indeed, until very recently HKMTRC has been regarded not just as an excellent transport organisation, but an excellent corporation. A new organisation successfully invented itself, and that excellence lived on. Not only had the processes and procedures that it developed stood the corporation in good stead throughout the tenure of several different chairmen, so that it managed the enormous growth of the system well, these processes and procedures have also transferred to other new organisations such as the HK airport development and, I like to think, have slowly percolated into the UK.

Sadly that reputation has faltered somewhat with the development of the Hong Kong Section of the Guangzhou–Shenzen–Hong Kong Express Rail Link, with major time and cost overruns.

## 5.6 Securing the Go-ahead

In September 1975 the project was approved, the Corporation was formed with two executive (Thompson and myself) and six non-executive Board members (including Sir Yuet-keung Kan, Sir YK Pao, and Mike Sandberg). Kan had been a Senior Official on the Legislative Council and was now the Senior Unofficial on the Executive Council. Pao was the biggest ship-owner in the world. Sandberg was Deputy Chairman, and subsequently Chairman, of HSBC.

The first item of the first Board meeting on 2 October was to confirm my appointment as MD, to which the Governor 'had given his approval.' Apart from general introductory business — standing orders, appointment

**PHOTO 5.2.** **HKMTRC Board members — Oswald Cheung, Mike Sandberg, Norman Thompson, Tony Ridley, Sir YK Pao.**

of auditors, banking arrangements and declaration of interests, the most important item was the letting of the first three civil engineering contracts.

The paper (MTRC Board 75/5), which I presented to the Board, said,

> "Contracts 201, 202 and 203 comprise the first round of international civil engineering contracts, the other tenders received to date being those from local contractors. At the present time the second round of civil engineering contracts, seven in total, is out to tender. These seven contracts contain the most difficult and costly parts of the civil engineering works whose total value is approximately HK $1,400 million out of a total estimated cost for all civil engineering works of approximately HK $2,0000 million at 1975 prices. It is most desirable that the competition between the international contractors is maintained during the period of tendering for the second round, tenders for which will be received in late December 1975 and early January 1976. Clearly, tenderers for the second round will be influenced by the manner in which the first round is awarded."

"It is postulated that in the event each of the first three contracts were awarded to Japanese contractors, then the greater interest recently shown by European and North American contractors would largely disappear. Equally it would be undesirable to fail to award any contracts to Japanese contractors because they might conclude that this was a direct result of the withdrawal of the earlier Japanese consortium bid and so decide not to tender for future contracts. Similarly the Japanese Ministry of Finance and Japanese commercial banks could well decide not to support Manufacturers Hanover's US dollar stand-by facility."

"The results of the tenders indicate that the lowest combination of contracts, excluding allowance for finance costs and exchange risk, is from Japanese contractors. However, if an award is made on that basis, there is little doubt that many European contractors would withdraw from current tendering, which could well result in Japanese contractors increasing their prices for the second round of international contracts. It is therefore recommended that, in order to maintain this competition, a premium is paid at this initial stage by awarding contracts both to European and Japanese contractors. It should be noted that this is the only occasion on which payment of such a premium is necessary because the award of the second round of civil engineering contracts will be made after all other tenders have been submitted and greater consideration can be taken then of the lowest possible price."

"The total price of these three contracts is approximately HK $8 million in excess of the original estimates. It is considered that the additional cost which might arise as a lack of adequate competition between the tenderers for the second round could greatly exceed this figure and might be of the order of 10–20 percent above the estimates."

What we did was to spread the contracts around, which I consider to have been one of the wisest decisions that the Board took during my time. Lesson — Accepting the lowest price bid for a single contract can be counter-productive.

The first contracts were signed later in the month but, because of recession in Hong Kong from mid-1974 to mid-1975, inflation was effectively zero. The new cost estimate, upon which new cash flow calculations were based, was HK $5,200 million since the escalation assumption of seven percent p.a. commenced a year later than previously. However, the prudent precaution was taken of assuming no more than that the previously stated cost of HK $5,850 million of the MIS was confirmed. Public statements, in fact, used a figure of HK $5,800 million. This, finally, was the figure which later allowed us to say that we finished the MIS 'on budget'.

### 5.6.1 E&M Contracts

Another wise decision was the 'grouping' of a number of E&M contracts. When the decision was taken to build the reduced MIS, it was also decided to do so on a multi-contract basis (25 Civil and 10 E&M). Of the E&M contracts by far the most expensive, and prestigious, was for the rolling-stock. Indeed there was a tendency to think that if the trains were German, then the system would be regarded as a German system, if Japanese then similarly. Consequently, governments, Export Credit organisations and bankers became deeply involved in the process, as well as the contractors themselves, and the competition was fierce (MTRPA (75) 81). The E&M contracts to be let initially were E1 (Rolling Stock), E2 (Signalling and Control), E3 (Telecommunications), E4 (Power Supply), E5 (Escalators and Lifts), E6 (Automatic Fare Collection Equipment), E8 (Station and Tunnel Auxiliary Equipment), and E9 (Kowloon Bay Depot Equipment).

The British team was led by Tony Sansome, the MD of Metro-Cammell, the potential rolling stock supplier. It quickly became apparent that a case could be made for groups of contracts (not necessarily the same ones) to be awarded to British, French, German or Japanese contractors. Each contract would be contractually separate but it was clear that if, say, four or five contracts could be coordinated in some way by a national team, then that would be advantageous to MTR. The Brits had one possible advantage in this respect, namely that Met-Cam had already tested their trains for Tyne and Wear on the test track at Longbenton, for which I had obtained government grant while I was DG there. They proposed to test the first Hong Kong cars there. Indeed, subsequent to them winning the bid, I received a beautiful photograph of the HK cars running on the test-track in Northumberland — in the snow.

In addition to careful analysis of bids for individual contracts, I began a series of discussions with the rolling stock bidders from all four nations. Eventually, Sansome proposed a deal whereby the Brits would 'co-ordinate' interfaces between the contractors for E1 (Metro-Cammell), E2 (Westinghouse),

**PHOTO 5.3.    HK Train in the Snow in Longbenton, Northumberland.**

E4 (GEC Rectifiers), E8 (GEC Hong Kong) and E9 (Vickers). The cost would be HK $30 million (about £3 million at that time). I discussed this with Thompson who was not impressed, so we agreed to seek a reduction and I went back to Sansome to try again. Eventually he offered, if they won each of the contracts, a support services agreement (known as E11) for the princely sum of HK $100 (about £10). The individual MDs never quite forgave us, but they knew that it had led to them winning the individual contracts. One long-term side benefit for me was that it first introduced me to Mike Nichols, who was given the coordinator role reporting to Sansome.

The late Mike Nichols, later chairman of the Nichols Group, became a friend over many years and reappeared when I was with LT and Eurotunnel, as well as being a colleague as a Director of the Major Projects Association (MPA), Chair of the Association for Project Management (APM), and deeply involved in the work on Risk we did on behalf of ICE with the actuaries.

Every three months the top brass of each company appeared together for two days discussion with our E&M team, and then had to appear in front of Thompson and myself, to account for themselves. No individual company enjoyed being 'hauled over the coals' in those circumstances (MTR Contract E11 Papers).

## 5.7 Integrating the MTR with the Public Transport System

On 18 August 1974, the meeting of government's Financial Management Committee had received a paper on, 'MIS — Financial Appraisal', that had attached to it the 'MTR Revenue — a Background Note' of 15 August (Ridley T, 1975). It said, inter alia, that 'in the early part of 1975 it appeared that the Revenue Review Study and the Comprehensive Transport Study (CTS) were producing markedly different results. It was feared that differences in modelling procedures was the reason. A great deal of comparative analysis has since been carried out and it is apparent that, while no two studies produce identical results, the philosophies of the two procedures are essentially the same. The main differences are now seen to relate basically to *input assumptions.*'

Midway through the construction period it was timely to concentrate some of my attention on the issues arising from the opening of the first stage of the MIS. I produced a paper on 'a complementary bus system' (Ridley T, 1977b). It started with two quotations, meant to concentrate government's attention. The first was from a Hong Kong bus company executive in April 1976, 'We are now working on a seven-year plan, bringing into the fleet about a 100 high-capacity double deck buses. We've decided to plunge ahead and paddle our own canoe because there is no overall government plan for transportation.' The second was from an October 1976 article about the Bay Area Rapid Transit (BART) system in California, 'Oddly enough, although the designers have always been explicit about the critical role of bus access to the system, the planning process was rather nonchalant about creating it. There was little effective planning for feeder bus services until the period just preceding opening day, and it is still quite inadequate in the outlying suburbs.' And, of course, I was still very mindful of the fully integrated rail-bus system designed for Tyne and Wear.

The paper said,

"A metro system has been recommended for Hong Kong because it is the only mode capable of handling satisfactorily the very large passenger volumes predicted in the major corridors. Daily volumes in excess of one million are projected for the MIS in the 1980s and government's Comprehensive Transport Study (CTS) indicates that in 1986 the MIS could accommodate about 20 percent of the total public transport demand. Clearly the opening of the MIS in 1979/1980 will have major impact on travel patterns and on the use of the existing travel modes. At present, however, little or no specific work has been done to design the future over-all transport network."

Government's relationship with the various public transport operators varied considerably, from the KCR at one extreme to PLBs at the other. MTRC, formed in September 1975, was charged with conducting its affairs on a commercial basis. The MIS was given the go-ahead on the basis that it must 'pay for itself'. Furthermore, vast sums of money for its construction had been raised on this assumption, albeit with the backing of government guarantees. The ability of the project to 'pay for itself' depended on two sets of factors. First were the internal factors including the efficiency of the design, construction and operational procedures. Second were the external factors, not controlled by the Corporation, which determined the conditions under which the system was required to operate. For example, if the Corporation were prevented from setting its fare levels commercially then its chances of paying its way would be severely hampered.

### 5.7.1 Complementary Bus Services

It was in the interest of the public and government, as well as of the Corporation, that the bus and PLB services in the area served by the MTR should be so designed as to complement the services it would provide. There were many transport arguments for such a policy, though clearly they were the prerogative of government. The paper was more directly concerned with the fact that such a policy was implicit in the various studies which preceded the decision to go ahead with the MTR and with the financial implications of adopting, or not, such a policy. It was important to start to define a transport network to include the MIS well in advance of its opening because of the complexity of the task and the necessity of gaining acceptance by many concerned parties to a new transport network. Also, some of the actions that could only be taken after a scheme had been agreed, such as the determination of bus fleet sizes and therefore new bus purchases and the planning if interchange facilities, required a long lead time. Finally, it was also timely to consider the policy issue because, with the publication of the CTS, government would then produce a Transport White Paper. The issue of a complementary bus system had therefore to be faced.

The paper considered, in turn, previous transport studies in Hong Kong; the possible financial consequences of a non-complementary system; the characteristics of a complementary system, including the consequences for interchange planning; and implementation. Discussions with government eventually led to their DIPTRANS study (Development of an Integrated Public Transport System) in 1978.

On 8 March 1978, government wrote to four firms of transport consultants, inviting them to submit proposals to develop an integrated public transport system that could be implemented 'as the MTR and an electrified Kowloon-Canton Railway (KCR) begin to materialise.' Quite soon Halcrow Fox and Associates were appointed and their team was to be led by David Silcock, author of the earlier RRS.

### 5.7.2 Emerging Controversy about Sharing Information

Thompson wrote on 17 October to Derek Jones, Secretary for the Environment and the boss of Richard Butler who was the chairman of the SG,

> "Tony Ridley keeps me informed of DIPTRANS progress and I am becoming increasingly concerned at the proposed methods to be used in this study in that information relating to revenue and average fares of the various modes of transport including MTR is to be circulated to the various operators albeit on a confidential basis. ... By reason of the provisions of the MTR Ordinance members of the Corporation Board are very conscious of the responsibility that they bear for viability as non-viability would have considerable financial repercussions for Hong Kong, bearing in mind the considerable maximum debt of HK $10 billion in 1982. ... My concern is that circulation of possible fare levels and numbers of trips per day on the MTR to other transport operators could well inhibit the Board in making its decision early next year as to the initial fares for the system. It is essential that the Board is not influenced in any way in arriving at this decision, bearing in mind that they will be responsible for the consequences of that decision. There is every indication that a conservative course should be

followed which will not necessarily please third parties who do not bear that responsibility."

Then, in a memo to Butler on 27 November 1978, Silcock said that initial DIPTRANS results suggested that it was not possible to develop an integrated system which gave 1 million passenger boardings at MTRC fares (HK $1.00–2.50 at 1975 prices) without either unrealistically high fares on non-rail modes, unrealistically severe network constraints on non-rail modes, or an unlikely combination of other input parameters occurring in practice.

But on 9 January 1979 Peter Miller, the new Assistant Commissioner for Transport spoke to the Hong Kong Section of CIT about 'Co-ordination and Integration of Public Transport'. He concluded with the 'gypsy's warning' that 'co-ordination and integration of public transport is likely to be the hottest transport topic during the Year of the Goat.' On 29 March his boss, Armstrong-Wright, wrote to Butler,

> "We do not yet have an acceptable network. With so many people who will be worse off, and an even larger number who will reap no benefit and will pay much more, and with so few who will directly benefit, I am certain that there will be massive public dissatisfaction and extreme implementation difficulties ... I am not sure how you intend to play this at the SG meeting but, in my mind, it would be quite wrong to pretend that we have network which is acceptable."

### 5.7.3 MTR Fares

Silcock's parting shot before leaving HFA had been in a memo to Butler and Armstrong-Wright on 13 January 1979, in which he said that 'the fare levels proposed by MTRC were too high to result in one million daily boardings.'

On 3 April 1979, Thompson wrote to Derek Jones, Secretary for the Environment, on initial fares for the Initial System,

> "The decision was made in the light of the need to advise the Executive Council prior to publication and having regard also to the time required by the Corporation for marketing, printing tickets and generally to make all necessary arrangements to allow opening to the public on 1 October. I also understand that you will be submitting your recommendations to the

Executive Council for the necessary restructuring of bus and PLB routes in the MTR corridors in August/September of this year. The Board, when making its decision, therefore had to assume such reshaping of bus and PLB services in order to provide the feeder service links to MTR stations and to change bus routes and PLB services in MTR corridors to complement the introduction of the MTR. Such a restructuring of bus and PLB services was anticipated at the time of making the decision to proceed with the MIS in 1975 and has been referred to in both the Executive Council and Legco and to the public generally. Indeed without such reshaping and movements in bus fares as mentioned later it will not be possible for the Corporation to achieve the revenue necessary to allow planned debt repayment as agreed by the government."

"I particularly emphasise this point as in the event of such reshaping not taking place or not being adequate, then the only alternative would be for the government to *subsidise* the Corporation's revenue which, to the best of my knowledge and belief, has never been contemplated. These particular points were noted by the Board as essential to its decision. I should also add that the initial fare structure of HK $1.00–3.00 with an average fare of HK $2.00 at 1980 prices also had regard to the agreed figure of some 850,000 passengers a day travelling on the MTR in 1981."

Thompson and I were then 'in attendance' at a meeting of the Executive Council on 8 May, following which he wrote to the Board,

"At the meeting of Executive Council held yesterday, it was agreed that the Corporation's fare proposals and the underlying assumptions be noted and, after the meeting I telexed/cabled you to the effect that the tentative meeting planned for 10 May would not be necessary. During the discussion that took place prior to the decision, it was apparent that members of the Council wished to ensure that members of the public would always have a choice between travelling on the MTR or some other mode of transport. In short, government did not want it to be suggested that bus services were being withdrawn to ensure adequate patronage and revenue for the MTR."

"Thus, although the need to provide bus feeder services to MTR stations coinciding with the opening of the railway was fully appreciated and agreed, there was considerable reluctance to withdraw bus services on parallel routes to the MTR until it had been demonstrated that there was a preference by some members of the public for the MTR which would then justify reducing the competing bus services. Previously it had been generally assumed that bus services operating on routes parallel to the railway would be withdrawn in part simultaneously and in part subsequently to the introduction of the MTR.' So far so good. The political role of the Executive Council was being properly exercised, but what Thompson wrote next in his letter had taken me aback when he said it to the Executive Council."

"In practice there should not be very much difference between the method proposed and assumed and therefore I accepted the position on behalf of the Corporation Board in the context of the Corporation's role to improve the public transport service. If desired, this can be discussed further at the Board's meeting in June."

I was not comfortable with 'not very much difference,' not least because I felt that the implications of decisions should be carefully explained to the Board and to the government.

I was very conscious of the fact that more major transport projects are undone by revenue forecasts than by finishing late and above budget, as became abundantly clear with the opening of the Channel Tunnel.

Lesson — Revenue forecasts are more likely to be erroneous than time and cost.

### 5.7.4 Press Interest

On 2 June, the fares issue broke in the press. 'Plan to boost MTR gets Exco veto,' said the front page headline in the *South China Morning Post* (SCMP). 'Exco has rejected proposals made by a firm of transport consultants which suggest that many bus services may be reduced or eliminated to force the public to use the MTR. The suggestions were made in a report of DIPTRANS by a firm of British transport consultants. Several of the proposals seem aimed at finding ways to forcibly encourage enough people to take the underground, in preference to other forms of transport, to make the MTR reach its revenue targets. The sweeping suggestions would have meant an increase in fares paid by a significant number of commuters.'

On 4 June, there was a further Board meeting. The minutes said, re fares, that 'members would recall the Board discussion on the subject when it was thought that the restructuring of bus services would be partly simultaneous and partly after MTR opening, whereas the Executive Council had agreed that any such restructuring should take place subsequent to the introduction of the MTR service. This position had been accepted by the Chairman on behalf of the Board as, in his view, this would not materially affect Corporation revenues so long as adequate bus feeder services were introduced coincidental with the opening of the MTR.'

## 5.8 Operations

When I went to Hong Kong in 1975 I had no particular idea of how long I might stay. Some expatriate members of the team stayed for the whole of the rest of their careers. Some went to Singapore or London Transport and then returned to Hong Kong. At the very least I was committed to seeing the constructing and delivering the first stage, the MIS, into successful operation. Although an engineer, I saw my management role as *delivering a transport service* and not just its infrastructure. Thus, in discussion with Norman Thompson, I determined to stay past the opening of an excellent service and thus became a proud member of what we call the *Hong Kong mafia*. I gave 12 months' notice in mid-1979.

### 5.8.1 Start of Revenue Operations

There were, in effect, two official openings for the MIS. A service began to operate from Choi Hung to Shek Kip Mei on 1 October 1979, with a First Train Ride (for top brass) attended by the Governor, Sir Murray MacLehose, on 30 September followed by a public Open Day starting at noon. The fares were at a premium, and not zero as might have been the case for an opening in a European city — and the proceeds were, in true Hong Kong fashion, given to the Community Chest. The Official Souvenir Book described the MIS and acknowledged the contribution of the contractors.

The *Asian Wall Street Journal* provided a foresight in its 24 September edition and the *Financial Times* had a big spread on 1 October. The *SCMP* talked about a 'trip into the future', and *Newcastle's Evening Chronicle*, with the heading 'Honourable Hong Kong metro run on time,' said that 'the man who masterminded Tyneside's metro today saw the birth of his brand new baby — the Hong Kong metro.'

### 5.8.2 Official Opening

The full-blown opening took place on 12 February 1980, when the service reached Hong Kong Island. This was a Royal Opening, many of us being presented to Princess Alexandra who was doing the honours at Chater station, with all and sundry taking part in the MTR's reception, which was followed by a celebratory lunch at the Hilton Hotel. Beyond that, the engineering of the Tsuen Wan Extension was going well and plans were being developed for decisions about a potential Island Line.

For the next while, after I had left, things went less well for MTRC. The attempt to finance the Island Line through property development foundered, and in November the Far East Economic Review was writing that, as HK's property market sank deeper into mire, an ambitious extension to the underground railway faced crisis. It spoke of rumours that developers Hang Lung were about to renege on its agreement with the MTRC, and said that Chairman Thompson had disregarded warnings about oversupply of office space on Hong Kong Island. This was followed in December by a statement that the HK government faced a scandal over financing for the latest extension of the MTR. By 1983 Wilfred Newton had succeeded Norman, and begun his 60-year stint as Chairman of MTRC, before moving to London Transport in 1989.

Of course, as Hong Kong always does, it overcame its problems and MTR went from strength to strength over many years.

### 5.9 A Timely Invitation

In 1979, the President and DG of CIT paid an official visit to Hong Kong. During conversations with them I was asked if I would deliver the next Overseas Lecture in London. I quickly said yes, and duly flew to London and delivered the paper, 'Hong Kong MIS — a Notable Achievement' (Ridley T, 1980a), of which an abridged version was subsequently published in CIT's journal Transport under the title, 'The Hong Kong Mass Movement Miracle' (Ridley T, 1980a). I have explained elsewhere that, in the midst of a large number of papers over the years, I have occasionally been presented with the opportunity of delivering a particularly significant statement to colleagues in my profession, which has justified me taking particular care with what I have written and said. Mass Movement Miracle was one such.

### 5.9.1 Where Does the Responsibility for Success Lie?

Fortuitously for me, the timing of my 1980 Lecture had been ideal from the point of view of re-introducing me to the profession in the UK shortly before I returned permanently. Jane would call it Jungian synchronicity. The success of the MTR 'adventure' in Hong Kong has been very well chronicled. One of the most important lessons that I learned arose from working with the Hong Kong government — that an intelligent and efficient client is very important, of which HK was the epitome.
Lesson — An intelligent and efficient client is very important.

Yet it was only many years later that another reason for our success emerged. I was with a former member of our team at a social occasion back in the UK — reminiscing about the past. "To what do you attribute our success," I asked him. "Oh, it's very simple," he replied. "You remember that once a month on a Friday evening we used to meet for a drink and an informal briefing by the Chairman and yourself. I remember very clearly you saying, "If and when we run into any problems with this project, I don't want any of you coming to me complaining about the contractors, or the consultants, or the bankers, or the government." I had concluded that, the task of the MTR project was NOT to deliver infrastructure, but to create a successfully operating transport service. "Never forget", I said, "that it is the people in this room who have the responsibility for delivering a successfully operating system." And, of course, they did.
Lesson — The task of the MTR project was NOT to deliver infrastructure, but to create a successfully operating transport service.

### 5.9.2 A Mature Successful Railway

Many years later *SCMP* published, 'Moving Experience: the MTR's First 36 Years' (SCMP, 2011). In it I am quoted several times, referring to the early days.' The MTR has been one of the great experiences of my life. ... Contractors from all over the world were efficient and welcomed the spirit of enterprise.' 'We had the pick of the world's expertise to work on this project. The flexibility of these engineers, as 'professional gypsies', was astounding. They flew in, found somewhere to live, and three days later began working on Nathan Road.'

At the time of the merger of MTRC with the Kowloon and Canton Railway (KCRC) I was called to Hong Kong as a railway expert when legislation on the merger was being deliberated. One of the things Legco was concerned about was the commercial freedom that the MTR had, which was part of the genius of Philip Haddon-Cave. These powers were beyond what normally happens to public bodies, such as the right to set fares. Some Legco members were keen to gain control, and freedom to oversee operational performance. I appeared and told them, "If it ain't broke, don't fix it. Be very careful with legislation which could tie the hands of the managers." In the event, Legco approved the merger with 30 out of 49 lawmakers in favour. However there was one final hurdle, to get the approval of MTR's minority shareholders, who approved by 82 percent in October 2007. On 2 December of that year the merger of the two companies took place.

## 5.10  Looking Ahead

### 5.10.1  Halcrow Fox and Associates

When it became clear that I was going to leave MTRC, Jim Tresidder suggested I might wish to join HFA when I got back to Britain, with a view to succeeding him after a period of time. I eventually decided to go in that direction, without giving any serious thought as to whether I would make a good manager of a transport consultancy. Through Jim, I was negotiating with Sir Alan Muir Wood (former senior partner of Halcrow, eminent tunnelling engineer and a Past President of the Institution of Civil Engineers) and Andrew Sharman (a fellow partner and Chairman of HFA).

At that time I retained a friendly relationship with Simon Murray of Legionnaire fame — an Englishman in the French Foreign Legion. Much more recently, for a time, he became the Chairman of Glencore. Murray was with Jardines, who was representing the Japanese in the second round of bidding for MTR E&M contracts. In 1978, he had attended a six week Senior Executive Programme (SEP), at Stanford University in California. He waxed lyrical about it, and it seemed to me that, if I could attend, it would suit me very well after five tough years on Tyneside, and five years in Hong Kong.

I struck a deal with Halcrow, whereby they would contribute towards my costs of returning home, with MTRC already committed to paying part of those costs. In addition, Halcrow agreed to finance the SEP, with course fee, accommodation and first-class return flights from London to California — quite a substantial sum.

However, before I left Hong Kong I was invited to Singapore and offered the role of CEO of their intended rail system. At some time during the building of the MIS we had had a visit from Prime Minister Lee Kwan Yu of Singapore. His government was considering whether they should build a transit system. Not surprisingly he thought it is worthwhile to come to learn any lesson from our experience to date in Hong Kong. He came with about six colleagues and used his time well during the day he spent with us as he cross-examined us on every aspect of our project from construction to financing. One could see how he had developed such a powerful reputation. Of course, it was important to him to reach a sound decision about whether to proceed because a transit system in Singapore was an even greater gamble than it had been to Hong Kong, with only half the population. It was also much less wealthy in the late 1970s than it is now.

He obviously satisfied himself because Singapore eventually went ahead to develop a system, though not without a number of fierce debates. Typically, Singapore did not develop an MTR, but an Mass Rapid Transit (MRT), perhaps in order to be different from Hong Kong.

Jane and I thought long and hard about the CEO offer, but eventually decided against. So I went ahead with the plan to join Halcrow Fox, and confirmed that in writing on 1 March. It crossed my mind that I might never be invited back. I have, in fact, subsequently paid many visits to Singapore. I returned as an Advisor to a Halcrow study for a North-East extension to the MRT. I was an external examiner to the Transport masters' degree at Nanyang Technological University, and subsequently held the SMRT Chair there for a short while. Finally and most recently, with others from Japan, South Korea, Australia, Belgium and the USA, I became an International Expert Advisor to the Minister of Transport for four years. I also had a close relationship with SMRT, as a result of forming the Railway Technology Strategy Centre, when I got to Imperial College (see Chapter 10).

### 5.10.2 Stanford

I had only been to Stanford once previously, for half a day when I was at Berkeley. It is totally different from Berkeley, in that it is very low density.

Indeed it is known, familiarly, as 'the farm.' My mere six weeks with the Senior Executive Program were fantastic and the (student) accommodation was modest, but adequate. However, the food was excellent and the lecturers superb, not only in their command of their subjects, but also in their delivery. There were nearly 200 people enrolled, from their late 30s to mid-50s in age, I being 47. About two-thirds were foreigners, i.e. non-North American. In every way it was a great refreshing and learning experience.
Lesson — A mid-career break can be very refreshing.

At the end of the programme we were all presented with a yearbook. It chronicled the staff, students, classes, exercise and the parties. It also included a series of jokes told by our lecturers. My contribution to the book was to assemble and edit them, 37 in all (Ridley T, 1980b). My introduction read, 'When we think back to summer 1980 at Stanford we will surely remember that our professors were each characters in their own right. Indeed, they made a good comedy team. Whether this was part of some subtle marketing technique is difficult to detect, but we will regard Stanford more warmly for the humour which went with the tuition.'

## Bibliography

Construction News (1978), *HK Mass Transit Railway*, Vol. 4, No. 5, May.

Edwards J *et al.* (1980), HK Mass Transit Railway MIS: System planning and multi-contract procedures, in *Proc. ICE, Part 1, Design and Construction*, Vol. 68, pp. 571–700, May.

Executive Council (1975), Memorandun (XCC(75)55), 9 September.

FF&A (1975), Revenue review study (Unpublished).

FF&P (1970), HK Mass Transit Further Studies, Final Report 1.

FMC (1974), Minutes, 13 June.

MTRPA (1974a), Minutes, 26 October.

MTRPA (1974b), Paper 75(24), Grants of Land.

MTRPA (1975a), Paper (75)54, Property Development.

MTRPA (1975b), Paper (75)71, Kowloon Bay Depot.

MTRC Board (75/5), Report on Tenders for Contracts Nos. 201, 202 and 203.

MTRC Board (75)81, E&M Tenderers.

MTRC, Contract E11, Support Services Agreement.

Ridley T (1975), MTR Revenue — A background note (unpublished).

Ridley T (1976), Mass transit comes to the tropics, Developing Railways 1976, in *Railway Gazette International*, pp. 25–28. (Reproduced as HKMTR, in *UITP Review*, Vol. 25, pp. 7–16).

Ridley T (1977a), Halfway there — and still on course, in *Railway Gazette International*, Vol. 134, pp. 9–13.

Ridley T (1977b), The relation between the metro and other public transport — a complementary bus system, Note (unpublished).

Ridley T (1978), MTR reaches the halfway stage, in *Hong Kong Engineer*, Vol. 6, No. 4, pp. 9–13.

Ridley T (1980a), The HK mass movement miracle, in *Journal of Computing and Information Technology*, Vol. 1, No. 3). (Abridged version of HK MIS — a notable achievement, Overseas Lecture, CIT, London).

Ridley T (1980b), Comic Relief, Stanford Executive Program, Stanford University, USA.

## Press

SCMP (2011), Moving experience: The MTR's first 36 years, SCMP Publishers, Hong Kong.

# 6

# London Transport 1980–1988

**How did London finally grasp the nettle of a declining metro system? How do you develop a programme for growth for one of the world's oldest transport systems in a great city?**

Some lessons

- Conflicts of interest need to be addressed.
- Project management 'by throat'.
- Make bold management appointments.
- The need for strategy in a successful business.
- Make explicit the nature and consequences of strategic planning choices.
- The 1980s were a crucial decade.
- Find support by learning from colleagues around the world.

## 6.1 Walking into a Challenging Environment

It is said that the end of a period at Business School — particularly in mid-career — can lead to a change of job, or wife/husband, or both, because this is a time for reflection about 'the rest of my life'. The change, for me, came only three weeks into my time at Stanford Business School in the summer of 1980.

### 6.1.1 How and Why

Ralph Bennett, Chairman of LT and one of my referees for the DG's job in Tyneside, telephoned in some distress. I knew that there was a major row between the Board of London Transport and their political masters in the

GLC, especially with the Conservative Leader, Sir Horace Cutler. Here is not the place to rehearse the row, but a short reminder is necessary to set the scene.

'Guilty,' screamed a headline in an evening newspaper, 'secret report condemns LT chiefs.' (*Evening News*, 17 June 1980) 'My team will stay, says defiant Bennett,' appeared on another page. 'What a way to run LT,' said another headline (*Evening Standard*, 17 June 1980), 'shell-shocked bosses blasted in secret report — out at last.'

In the *Daily Mail* next morning, 'Bus bosses just can't cope'. The headline piece, with a photo of Bennett (Chairman), Stansby (Deputy Chairman), Glendinning (finance), Quarmby (buses) and Cameron (Personnel), went on, 'the men who run London Transport were savaged yesterday in a report they commissioned themselves. The Board was described as self-satisfied, shell-shocked and weak in the skills required to run a large business, and indeed manage itself as a board' (*Daily Mail*, 18 June 1980). Other newspapers joined in.

This crisis had been developing in LT for at least two years. In 1978, John Stansby was appointed Deputy Chairman and had set in motion proposals for structural change, of LT management, with a paper on the 'Reorganisation of London Transport 1978–1979' (LTE, 7 December 1978). Ralph Bennett had been appointed Chairman of LT Executive in the previous March, with Michael Robbins and David Quarmby assuming their respective roles of MD of the Railways and Buses. Management Boards for both LT railways and LT buses were created, each with Operations, Development, Engineering and Finance Directors. Significant changes were also made to the organisation of the 'centre'.

In January 1979, Cutler appointed Leslie Chapman to the Executive. Chapman, a retired Civil Servant, who became known as the 'mad axeman,' by virtue of being photographed on his first day with an axe over his shoulder, thus signalling his intent. By November 1979, while the re-organisation was still under way, Chapman produced two scathing letters to Bennett and the GLC about manifest inefficiencies in LT, from wasteful and unnecessary use of cars and chauffeurs, and hospitality, to size of offices and shortcomings in the Works & Buildings Department. Deloittes were asked to comment on Chapman's allegations and LT were asked to respond to the Deloittes report. Then, in April 1980, LT were forwarded, by the GLC, a copy of Chapman's comments on that response. The row rumbled on and the battle between LT and the GLC was not going to go away, with Cutler and the GLC aided by Chapman, and (perhaps) Stansby.

At some stage Bennett decided or was persuaded into appointing a consulting firm to carry out a review of the operation of the Board of LTE. Its report was devastating and for some time Bennett resisted making it public. But Cutler started to demand its publication.

I was in bed in my room at Stanford when the phone rang. "Tony," said Ralph "I want your help." "What can I do for you Ralph?" "I want you to be the Managing Director of London Underground." Didn't he already have a Managing Director of the Underground (Bill Maxwell)? Yes he did, but he would take care of that. "Well," I said, "you know that I have only just started a new job with Halcrow and I am going to become MD of Halcrow Fox." Then the fateful question, the same question that had been put to me about Hong Kong by phone in 1974, "Are you prepared to talk about it?"

There was to be a meeting of the GLC in two weeks' time when, because of the poor financial state of LT, hefty fares increase would be adopted, which Cutler would not enjoy. We agreed that I would fly home to meet Ralph. We had a long discussion prior to a meeting with Horace Cutler the next morning. The job was mine and Horace would be delighted, as would his colleagues, if I would take the job. I had agreed with Ralph that I would raise his position with Cutler. "It is important to know who is my boss." He mumbled something like "nothing precipitate will be done." That, however tenuous, seemed to please Ralph.

It was now necessary to discuss the matter with the Halcrow partners, Alan Muir-Wood and Andrew Sharman, who had appointed me. Virtually everyone in the firm had worked their way up the tree. Not everyone in the firm thought I was as good a catch as Alan and Andrew seemed to be. For me to leave the firm three months after joining was not exactly something that they were going to enjoy.

Both, however, were splendid and understood my predicament. I may have been helped by the fact that Halcrow together with Mott MacDonald, both first-class tunnelling firms, had jointly been LT retained consultants for many years. As such they did not want to run foul of a principal client. It was quickly agreed that I would not leave them but would be seconded to LT, on the basis of 90 percent of my time, for a period of three years. But I would remain 'their man'.

A principal concern of Halcrow was the 'leaning over backwards syndrome.' If I, seconded from them to LU, were to appoint a tunnelling consultant for a job, would that be regarded as a conflict of interest and preclude them. That matter was handled in a paper on 24 July to the Policy and Resources and Planning and Communications Committees of GLC, recommending my appointment. In order to protect the position of Halcrow, it was understood by all concerned that, in accepting, I would continue my association with HFA and Halcrow. I would, in all circumstances, be acting in a professional capacity and would be identified with the organisation which had laid upon me the duties I had assumed. It was not anticipated that this would involve any conflict of loyalties or interests since both Halcrow and HFA were consultancy organisations with no commercial constraints, and they would only consent to perform duties or to instruct their staff to perform duties that were consistent with the maintenance of strict codes of professional ethics. It was agreed that I, or LT, would not be precluded from making such appointments of Halcrow or HFA as seemed appropriate.

In the event that decisions had to be made on which it might be considered, or alleged, that some conflict of interest arose between the different responsibilities that I was assuming, it was agreed that the matter should be referred immediately to the Senior Partner of Halcrow, the Chairman of HFA, the Chairman of LT, and the DG of GLC. This was just one of the occasions in my career that a possible conflict had to be addressed. They do not cause difficulties, as long as they are foreseen and acted upon up-front. No problem arose throughout my term of office, although my friends in Motts, who had done so much to help me with the Metro in Newcastle, were not best pleased by my appointment. Halcrow and Motts, as LT's tunnelling consultants, no doubt felt competitive with each other.
Lesson — Conflicts of interest need to be addressed.

Shortly before the end of my three year stint, I went to Keith Bright, by then Chairman of LT, and suggested that the secondment should be extended by two years. "Nonsense," he said, "we will appoint you to a full five year term." The prospect of being seconded for a total of eight years was clearly so ludicrous to all parties that I left Halcrow and became a full-time LT employee. I did, however, become a Non-Executive Board Member of Halcrow Fox, which role lasted for a quarter of a century.

### 6.1.2 Networks of Relationships

Perhaps, the idea of my original appointment was not Bennett's alone. Sometime before I was due to return to Britain, Jane and I had hosted a visit to the MTR by both GLC and LU personnel. Correspondence suggests that they were very impressed. Charles Cope, Operations Director, was in the party and no doubt contributed to the view that MTRC was a well-managed organisation (Cope C to T Ridley, 20 May 1979).

In a letter of thanks the GLC Chairman said, 'This opportunity to see for ourselves what you are achieving in Hong Kong was one of the main aims of the tour and was amply justified. I was both thrilled and impressed not only by the vigorous plans that you have made but also by the dynamic way in which you are now implementing them. ... And to you and Jane our further thanks for the delightful dinner you so kindly gave in your flat. It was both a most useful and enjoyable occasion' (Mote H to T Ridley, 23 February 1979).

After accepting an offer from Cutler, I flew back to rejoin the Stanford Programme. The appointment was to be confirmed by the GLC on 8 July, together with the fares decision. On the morning of 7 July at 7.30 am (3.30 pm in London) I received a telephone call. "Tony, this is Ralph. I think you ought to know that I am going to be fired tomorrow." I expressed my surprise and asked what he thought I should do. He thought that I should 'think hard.' There was no decision as to his successor. To help my position I rang Jim Swaffield, the DG of the GLC. I told him that I could not be bound by my job acceptance if I (or he) did not know who the new LT Chairman would be. He said that there was no need to reach any immediate decision. The job offer would stand. By the time I returned to London four weeks later, they would have appointed a new Chairman and I could make my mind up after that. Although complicated, it seemed as though I could not lose, except for the possible embarrassment of 'rejoining' Halcrow.

Bennett issued an LTE staff notice (25 July 1980) announcing his relinquishing of the chairmanship of the LT Executive and that he would leave LT; the standing down, for health reasons, of Bill Maxwell as Managing Director (Railways) and the intention of GLC to appoint me as MD (Railways), and a member of the LT Executive from 1 September; the retirement as a part-time member of the Executive of Michael Robbins, a former MD (Railways); the extension of Sir Peter Masefield's part-time membership of the Executive from 1 January 1981 for a further year — a position he had held since 1 January 1976; and, finally, the renewal of Leslie Chapman's part-time membership of LTE for a further two years from 1 January 1981.

Shortly after returning to London I had lunch with Sir Peter Masefield, a part-time, Non-Executive Member of the Board since 1973, who had been appointed Chairman for an interim period of two years. He had aviation in his blood, but had started as a technical journalist. He went on to work for Beaverbrook during the war, served in the British Embassy in Washington and worked in the Ministry of Civil Aviation. Then he became CEO of BEA, MD of Bristol Aircraft, MD of Beagle Aircraft, and then Chairman of BAA. I very quickly confirmed that I would join him and LT.

Masefield was just about the most polite person of my lifetime. When we met Ken Livingstone and Dave Wetzel, after their 'bloodless *coup d'etat*,' he was charm personified, the ultimate gentleman and the ultimate democrat. Nearly all of his business and personal letters ended, 'with the warmest of good wishes.' But he was also a stickler for good English — ever the journalist. How many Board chairmen would issue a missive to Board members and senior staff, headed Accuracy of Expression. 'Everyone evolves his, or her, own writing style over the years, for better or for worse. Individual styles develop and, within limits, they are a Good Thing. No one should be forced to conform to a rigid pattern. Even so, good English (in the broadest sense) is important to both clarity and accuracy of expression which, after all, is the purpose of setting things down on paper.' Then he gave a list of proper and not proper usage — the word 'it' often leads to long-winded and misleading statements; split infinitives are rank bad English; do not confuse 'prevaricate' and 'procrastinate,' 'flaunt' and 'flout'; there is no such word as 'alright'; 'decimate' means reducing 'by one-tenth,' not 'to one-tenth' and so on and so on.

After conversations with David Quarmby, MD (Buses), who I knew from GLC days, and Ian Philips, the new Finance Director, I had not taken long to decide it would be good to work with Peter.

The *New Civil Engineer* (NCE, 1980) greeted my arrival with an interview, together with my picture on the front page of the 20 November 1980 edition and the title, 'Ridley takes the Tube'.

"Tinkering with the controls of one of the world's most complex train-sets may be the private fantasy of many civil engineers, but few would envy the problems confronting Dr Tony Ridley who took charge of LU's network in

September. Though financially somewhat better off than London Transport's other big people mover, the buses, the Underground has shared equally in the barrage of morale-sapping criticism directed at LT by both press and public in recent years. It was as much an attempt to restore morale as to improve operating efficiency that brought Ridley to LT as MD."

"The occasion was preceded by a blaze of soccer-style headlines making much of the fact that LT's political masters, the GLC, had agreed a 'transfer fee' (of £60,000) with transport consultants Halcrow Fox and Associates for the services, four and a half days per week of one of its joint MDs. Though flattered by all the attention, Ridley is keen to put the deal in perspective. The figures are right but it was not a transfer fee of the kind people normally associate with that term. I had only come back to Britain in April and Halcrow had invested money in me, not least for two months at Stanford. It was hardly appropriate to turn my back on them. So I retained my links with the firm, giving them one-tenth of my time, and the so-called transfer fee is to pay-off the front end costs they had invested in me and to pay my pension, for example, which stayed with Halcrow and which LT is relieved of."

"In all the publicity his new appointment inspired, never once were his credentials for the job questioned. It is difficult to argue with a CV that encompasses high-level responsibility for two of the most successful transport ventures of recent years, the Hong Kong and Tyneside metros. Ridley was MD of HKMTRC from February 1975, when its first construction contracts were being offered, until his return to Britain early this year. Now it is carrying 600,000 passengers per day. But he confides he may never have taken part in this major success story were it not for a change of heart back in 1974. Then happily ensconced in his native Northeast of England, and having successfully steered the T&W metro from conception to the early stages of construction, he could see little to recommend uprooting to the Crown Colony."

"I originally rejected the idea, seeing it only as a massive construction project. But I then discovered it was much, much more than that. We had big organisational and managerial problems of raising money and of getting personnel together in Hong Kong in a very short period of time. And training local staff who had never before seen an underground railway, but who are today running the operation, was an achievement just as important as building it on time and within budget."

"Ridley's part in the development of Newcastle's metro, though truncated by his move to Hong Kong, is legendary — and can still bring grimaces from the transport executives of Manchester and other cities who might have felt that they had prior claims to government funding. The Metro was not thought about until the end of 1971 and we were already digging the first tunnels in 1974", he says. "In fact it opened to operation only nine years after the idea was conceived. I know of no other urban railway in the world that has gone as fast as that. Even Hong Kong, of which I am very proud, opened to operation some 20 years after it was first proposed."

NCE concluded by saying that, still a youthful 48, Ridley was even then beginning to map out his future beyond the three-year stint offered by LT. "It has always been dangerous to forecast my life, it's taken some strange turns. But my guess is that London will be my base for the rest of my life, though I didn't say London Transport!" And so it has turned out — I did stay in London, but there were some complex turns along the way.

### 6.1.3 Early Challenges

I started with LT in September. I had a somewhat sobering experience on my second morning. I was due to meet with John Stansby and several other colleagues in his office at 9 am. John and I were the only two there at the due hour, but others wandered in one by one. At 9.20, when all the attendees had still not arrived I asked John, 'is this really the way things are done in this place?' I could not believe it — and it certainly was not Hong Kong.

Another shock followed. It transpired that two of my Board colleagues were renowned for being, quite often, not capable of dealing with business after lunch. People simply left matters until the following morning when they had 'recovered,' but nobody had thought it was necessary to confront the reason for the problem. The LT Senior Officers' dining room was also known to be a 'good perk' and drinking at lunchtime the norm. In those days railway dining rooms had not moved to being 'dry,' not in order to prevent over-imbibing but to bring senior management into line with operating and maintenance staff, for whom drinking while on duty was firmly forbidden. Nonetheless, the Board, at some stage, closed the Senior Officers' dining room, rather than turn it 'dry'. To our dismay it also killed off, at a stroke, valuable informal communication across the organisation. Not entirely what we wanted!

### 6.1.4  Put in My Place?

After three days I was due to chair my first monthly management meeting
with all of the Directors (Charles Cope, Operations; John Cope, HR; Leslie
Lawrence, Engineering; Bob Dorey, Development; and Geoff Allen, Finance).
The start was delayed because, after the fracas between GLC and LT, Alan
Greengross the Chairman of the GLC Transport Committee had made some
unfortunate, slightly critical remark about the new Board under Peter
Masefield. The press jumped on it and Alan, a nice man, recognised his error
and a 'photo-op' was quickly arranged between Peter and Alan on the plat-
form at St James Park station. The new MD was invited to accompany them.
When I eventually joined the management meeting all eyes were on me.
I apologised for my unfortunate start. The first person to speak was Charles
Cope, an excellent Operations Director, whom I had met before I went to
Hong Kong, and when he later visited Hong Kong. He, politely but point-
edly said that he was not used to the MD going on the railway unless accom-
panied by himself. This I would change.

LU was full of splendid people, of many years of devoted service, but with
one fatal flaw. They all seemed to believe that nobody could tell them anything about managing the Tube unless they had served at least 30 years —
including, in particular, the new MD recently arrived from Hong Kong! It was
immediately apparent that the challenge was very different from the MTR.
A trip to Hamburg was to be part of the culture change.

## 6.2  Engineering and Project Management

Another challenge was to address the role of project management within
engineering. In 1980, the discipline was only just developing in Britain, but
I was still surprised by my first experience of it, or lack of it, in the
Underground. At every monthly management meeting the last items on the
agenda examined projects that had got into difficulty. At my first meeting,
after three days, there were four. I let two proceed, directed certain actions
on a third, and held up one for further feedback. A month later something
very similar happened again, so I asked, "How many projects am I responsible for?" "Define a project, MD." "Well, something self-contained and costing more than £50,000." Remember that this was 1980 when that sum was
worth much more than today.

A month later I was told, "MD, you are responsible for 632 projects." "Thank you. Now give me a list of the individuals who are personally responsible for each one of them." Dropped jaws, lowered heads. Nobody spoke. "Come on, who is responsible for this one?" "Ah, that is the responsibility of the Chief Civil Engineer, the Chief Signalling Engineer and the architect." I thereupon announced a new paradigm — 'management by throat'. "Whose throat do I grab when something goes wrong?" I immediately appointed Mike Nichols to help me to develop some basic processes. Very crude, but my new colleagues did start to recognise, from the outset, the need for proper project management.

Lesson — Project management 'by throat'.

Civil engineering was also another area where I made an impact, albeit unknowingly. The Chief Civil Engineer announced that he was about to retire, after many years of worthy service. A competition was arranged, with interviews. I had been tipped off by Quarmby and Phillips about some of the candidates. Cliff Bonnett was an excellent appointment, but the shock to the system was that I did not appoint the sitting Deputy. It was clear he would be disappointed, but I had no idea what a strong message (the best person for the job, not buggins' turn) the appointment would send through the organisation.

Lesson — Make bold management appointments.

## 6.3 Strategy

Stanford had stressed the importance of the role of strategy in the development of a successful business, a principle I had understood in Newcastle. The Joint Policy Statement that we developed in the first year had been of inestimable value as we pursued our subsequent plans. In Hong Kong I was little involved in strategy — the emphasis was on delivery — but the HK government had developed its proposal for the MTR on a sound basis, including plans for the future extension of the system.

The reorganisation taking place within LT meant that the Rail and Bus businesses were committed for the first time to producing their own Business plans, for submission to the Executive and thence to the GLC.

The (draft) Strategic Plan for the Rail Business 1981 was produced by the end of March 1981, under the leadership of Bob Dorey, the Development Director (LU, 1981).

In the introduction it had a series of telling but worrying statements. 'In recent years the (LT) Executive has presented to the GLC in May a 10-year capital programme and a 5-year revenue plan. The capital programme has largely been the result of programmes produced within the businesses, but the revenue plan has been initiated from the centre with some limited involvement by the businesses — usually confined to the Finance and Development Directors, who have had little influence over the final document. The most recent five-year plan was debated by the Executive for about *5 minutes*. Against this background, it is hardly surprising that fundamental strategic issues have not been tackled and that line managers, who have a major contribution to make, have had no sense of committing to previous plans.'

While the words were worrying, at least there would be one ally on the new team. The introduction described the new approach adopted, whereby businesses were producing their own strategic plans, covering both capital and revenue issues. An effort had been made to involve managers as far down the line as possible.

It went on to say that not all of the questions that should be answered had been in the document, or even raised. But planning was a continuing process and the work that had contributed to the production of the plan was just the start of a process and one that should eventually involve management and staff at many levels.

### 6.3.1 Developing a Strategy for the Rail Business

Meanwhile, we appointed Booz Allen to review the new organisation and to report on the Development of a Strategic Planning Process for the Rail business. A progress report at the end of April 1981 (Allen, 1981a) said, 'LT Rail's capital investment plans are now much less ambitious than those proposed in the 1970 programme. Annual capital investment programmes going forward, although substantial, are dominated by renewals expenditure to maintain present basic levels of safety and service. LT Rail's business choices are highly dependent on decisions taken by the GLC and the LT Executive'.

The report described the conceptual approach to planning for the Rail business. The prime purpose of strategic planning is to make explicit the nature and consequences of strategic choices. If no real choices exist, planning focuses on tactics and operational management. It went on to define alternative major business options involving choices on the long-term cost structure of the railway. Thus was strategic planning launched.

Lesson — The need for strategy in a successful business.

The current process did not address alternative roles for the railway — only marginal changes within the current role. The 1986 time horizon for revenue plans impeded evaluation of some major capital projects aimed at long-term cost reduction. Within a given role longer-term investment programmes had been designed by the engineering function and appeared to be driven primarily by identification of technological opportunities rather than cost reduction needs; and, within an assumed level of funding, short-term projects appeared to take precedence over major capital projects supportive of major business options.

The strength of the current Rail business planning process was in dealing adequately with many tactical options, and strategic alternative roles and business options could not be evaluated practically in the same framework as tactics. Overall, there was a dichotomy between Rail management's dissatisfaction and Group Planning's satisfaction with the planning process.

The final report, in June 1981 (Allen, 1981b) said,

> "The long-term strategic objectives for London Transport as a whole set the context for planning in the rail business. Options open to LT and to its individual businesses can exist at three levels. Level 1 choices involve alternative roles for LT and for its individual businesses in contributing to passenger transport in London. Level 2 choices involve major business options for the Rail business within a defined role, involving fundamental choices on cost structure, quality and employment. Level 3 choices involve tactics within a defined strategy. The essential strategic choices for LT (Rail) call for a 10–20 year planning horizon. Rail business planning is part of an integrated hierarchy of planning."

Lesson — Make explicit the nature and consequences of strategic planning choices.

The report described strengths and weaknesses of the current strategic planning process. The Rail business currently had a solid basis for planning at Level 3, but no explicit planning process for dealing with Level 2 strategic options. There was a high cost in the resulting lack of communication upwards and downwards within the LT Executive and LT Rail. There was a similar lack of an explicit planning process at Level 1 Board level. The report concluded with recommendations for a future planning process for the Rail Business. For the first time, within 12 months of my arrival, we could start — but only start — to decide the long-term direction of the business.

## 6.4  A New Administration for London

Throughout the 1980s LT experienced a series of events, each of which represented a major challenge, and buffeted us throughout my time with the Underground. First we had the arrival of 'Red' Ken Livingstone, followed by Fares Fair and its overturning by the House of Lords. There followed the appointment of Ken's political appointees to the Board of LT, and their subsequent removal by Nicholas Ridley, Secretary of State for Transport. In 1987, the disaster of the King's Cross fire was followed by a 93-day Public Inquiry.

Joining the GLC in 1965, appointed by a Labour administration, Brian Martin and I had become a small part of the subsequent election when Labour claimed to have 'reversed the brain-drain' by bringing us home from the States. It was in vain. The Tories won the election under Desmond Plummer.

In 1980, it was the Conservative Chairman of the GLC, Horace Cutler, who appointed me to the Board of LT. In May 1981, the next election was won by Labour. The implication of having two competing political masters meant that the Board of LT had written two policy documents — one with a blue cover and one red — in a pale imitation of the way that Civil Servants prepare for alternate governments at a national election. Labour was led by Andrew (later Lord) McIntosh. What we had not counted on was that there would be an immediate peaceful '*coup-d'etat*'. Within 12 hours we discovered that we had a totally different master in Ken Livingstone. His Chair of Transport was to be Dave ('yours for socialism' as he subsequently described himself) Wetzel.

'Red' Ken needs no introduction, having gone on to develop an international reputation as Mayor of London. But at that time he was comparatively unknown, certainly to us. He had already been active in local government, but 'taking over' the GLC began to put him on the map. Subsequently on his way to the Mayoralty, he not only made a mortal enemy of Mrs. Thatcher, he totally fell out with the Labour party whom he defeated as an Independent when he first ran for the position.

I had only met him once previously. When I became MD of the Underground and was asked to several speaking engagements to publicise the challenges of the job. One such invitation was to speak at a seminar at LSE. Ken was there, and remembered me well. As he told a group of acolytes when we met years later at a reception in Buckingham Palace hosted by the Queen to honour the London 2012 Olympic Bid Team, "I remember Tony, when he first came back from Hong Kong. He was speaking at LSE and he said, 'marvellous place Hong Kong, no bloody politicians". Not quite true but added Ken, "I thought, I'll keep an eye on you me lad". On that occasion Ken was very jovial, but there were ramifications later as he was determined not to allow me to become a Board Member of the Olympic Delivery Authority.

When the newly powerful Ken and Dave came to 55 Broadway to meet the Board it was wondrous to behold the studiously polite and charming Sir Peter Masefield receive them with his inimitable courtesy. There was a certain irony about Dave's presence. A former bus conductor, apparently he had been turned down for a job with London Buses only six weeks previously. A warm and friendly man, not like so many other left-wingers who seemed particularly humourless to me, he had an exotic political background. He called himself an anarcho-syndicalist, a group which research suggested, had died out with the Wobblies in Seattle in the 1920s. Many years later he left Labour for the Green Party.

## 6.5 Fares Fair

The first priority of the new administration was a significant reduction in fares. Central to the Labour manifesto, and therefore its programme when elected, was a reduction in bus and Tube fares of 25 percent. This was to be financed by a supplementary charge on the rates of 5p in the £. The initial discussions in the Labour Group were directed towards a zero-fares policy but, the unions representing the ticket collectors and bus conductors panned that idea, for obvious reasons (Carvel, 1984). Led by Peter Masefield, as good democrats, the Board of LT leapt into action and produced a proposal for how the fare reduction might be implemented. It was at this stage that Dave Wetzel's transport background first became valuable. Ken had wanted a 25 percent reduction in every fare, but LT worked out a better proposal for zonal fares, which happily led to the introduction of the excellent Travel card some years later. When the details were produced the overall reduction turned out to be 32 percent.

Introducing a zonal scheme, after a graduated fares scale, is politically difficult if not impossible if the average fare remains unchanged, since some fares go down but some go up. With a one-third reduction in average fares, no-one's fare need go up and some people benefit substantially. Dave argued successfully, within the Labour Group, for our proposal rather than the simple identical reduction from each fare. But, perhaps, we were too much good democrats. We did not point out, or even examine, the possible legal shortcomings of what was proposed. As good 'civil servants' we might have done but, in fact, it was the GLC Officers who were Ken's 'civil servants'.

### 6.5.1 Challenge by Bromley Council

The impact of the reduction worked like a dream and passenger numbers immediately began to rise, on both buses and Tube. But the Conservative leaders of Bromley Council decided to challenge the supplementary precept for the levying of rates to pay for the cost of the scheme. They sought a judicial review of the precept; the respondents to their application to the Divisional Court, on 3 November 1981, were the GLC coupled with LTE as the implementers of the fares reduction. Their grounds were that the precept was inequitable as between the property tax payers and the fare payers. They were particularly incensed on the grounds that, as someone said rather lightheartedly, 'Bromley had no Tube and precious few buses'. The Court dismissed Bromley's application. By 10 November, Bromley had appealed to the Court of Appeal, which found in Bromley's favour. This Court, which included Lord Denning as one of the three judges, declared the supplementary precept to be ultra vires and quashed it. But the GLC and LTE were given leave to appeal to the House of Lords (now the Supreme Court).

On 17 December, the five Law Lords unanimously ruled that the GLC's cheap fares policy and the supplementary rate were unlawful (Carvel, 1984). LT was immediately illegal, and charging illegal fares. Furthermore, the several tens of millions of £s received from the GLC from the time that the fares had been reduced, was also illegal. Thus, it was necessary, not only to return the fares to a 'legal' level, but also to repay the millions received from GLC. We sat in our board room distinctly bemused. In the USA, the Supreme Court gives a 'majority' judgement and, where necessary, a 'minority' judgement. In our case the five Law Lords had found unanimously against us, but there were five different written judgements explaining why. We thus had to get further legal opinion to form a view as to what might be legal!

It was not even certain whether any subsidy was legal. In one scenario, to put LT into a position of being without any subsidy at all would have required fares to be quadrupled, which clearly would have been ruinous. We eventually took a pragmatic decision. The fares were doubled, on the grounds that this would probably be legal and would not destroy the services. It was a no more profound decision than that.

Of course, the fares were not now twice the figure we had started with. Starting with 100 percent, then reducing by 32 percent, produced 68 percent. Doubling that arrived at 136 percent i.e. the fares were now 36 percent higher than we had started with. In his Lords judgement, Lord Scarman had said that it was plain that the 25 percent overall fares reduction was adopted not because any fares level was impracticable, but as an object of social and transport policy. It was not a reluctant yielding to economic necessity, but a policy of preference. In doing so the GLC abandoned business principles. That was a breach of duty owed to the ratepayers and wrong in law.

In 1983, the GLC tried again. This time they did not simply say that 25 percent seemed like a good idea. They produced a careful and detailed analysis of transport in London, and produced a new scheme for a further 25 percent reduction in the fares based on their carefully considered balanced plan (Forrester, 1985). Going back to the previous arithmetic — taking 25 percent off 136 percent produced 102 percent, back where we started! But because a coarser fare structure, necessary for the Travel card, was only politically possible with overall fares reductions, all of this mayhem did lead to a 'once in a lifetime' opportunity twice in two years. For a second time, we were able further to coarsen the fares scale.

## 6.6 A New Chairman

When Sir Peter Masefield agreed to take the Chair of LT in 1980, he did so for two years. He was succeeded as Chairman by Dr Keith Bright, an industrialist with a PhD in chemistry. He had worked in senior positions with Formica International, Sime Darby in Malaysia, and Associated Biscuit Manufacturers. He was also a Non-Executive Director of the British Airports Authority.

Somewhere through all the to-ings and fro-ings of Fares Fair, it must have struck the GLC that, when it came to facing the law, LT's policies were generally regarded as legitimate, but that GLC's often were not. A simple solution was obvious — appoint sympathetic non-executives to the LT Board. In July 1983 the GLC proposed to add six non-executives to our Board. Arthur Latham was effectively the most senior, having been a Borough Councillor for Romford (1952–1965), and for Havering (1964–1978) of which he was the Leader (1962–1965). He was an MP (1969–1979), and had become chairman of the Greater London Labour Party in 1977.

The Chairman was entitled to be consulted on proposed Board appointments and they each went through an 'interview' process with Bright. All survived, except one who was deemed to have an 'error' in his CV. Bright was, however, distinctly underwhelmed with the lack of Board skills and experience of several others. He put forward alternative names, but none of them commended themselves to the GLC, whose purpose was clearly to pack the Board with sympathisers. This they were entitled to do but, for a while, LT was turned into a political bear-pit. The announcement of the appointment of the five GLC nominees was made on 21 July at the same time the renewal of my appointment to the Board, for a period of five years, was announced.

Arthur Latham, as Senior Non-Executive, soon began to worry that, not having six non-executives on the Board, might regularly produce 5–5 splits on votes, which they frequently wanted to take. So GLC proposed to appoint Merle J Amory, a 25-year old Brent councillor, with trade union experience, and a school governor. Nonetheless, as she was a secretary by profession, Bright felt that too many of the non-executives were already insufficiently experienced in business or the complexities of a very large and costly operational business-like LT, so he reminded Ken that he had said in July that, 'if and when a further appointment was made to the Board, he would expect any such person to be such an obviously useful addition that he or she would have the support of all parties in the GLC and be welcomed by all the existing members of the Board.' Nonetheless, Amory was appointed.

We started off friendly enough but soon got into debates about which committees the new members could or could not attend, what was the dividing-line between Board and management matters, and whether, as they believed, information was being withheld from them. The next argument arose when, in November 1983, it was proposed that Latham should become a full-time, Non-Executive Board Member, a role with all kinds of opportunities for mischief. In the event, Latham was indeed appointed a full-time non-executive.

## 6.7  Implementing Strategy: The Need for Renewal and Improvement

Having produced a comprehensive new strategy in 1981, LU had an improved way of looking forward and handling our affairs, but the process was quickly overshadowed by the arrival of the 'new' GLC, Fares Fair, battles with new members of the Executive, struggles to improve our operating performance, and the introduction of One Person Operation. Nonetheless, the strategic process was much improved, and later came into its own when the increase in ridership of 60 percent over five years presented us with a new set of challenges.

Over the years I have regularly gone 'on the record' about my professional activity, not always to the liking of colleagues or bosses. However, I do believe that it is part of the job of 'the man at the top' to keep the public, professionals and staff well informed, a policy that paid massive dividends during the promotion of the T&W Metro. In July 1981, some 10 months after I had joined LT, I had the opportunity to lay out my stall when invited to give a talk in the Centenary Briefings series of the London Chamber of Commerce, on the subject of the Future of the Underground (Ridley T, 1981).

This began by saying that they had made a wise choice of subject, because the topic was of great importance to the Chamber.

"Transport is of great importance to any great city, a fact that has never been truly acknowledged in London. Equally the Underground — a reserved right-of-way through and under the urban complex — is the back-bone of the whole public transport system, together with BR commuter services."

"The fact of the centenary reminds us that London is both fortunate and unfortunate, in already having a vast Underground. The first Underground railway in the world came into operation in 1863, running from Paddington to Farringdon Street. As I speak the Railway Business puts into service up to 480 trains each weekday, running more than 30 million miles annually, serving nearly 280 stations over 255 route miles. It manages and maintains 249 stations, 699 miles of track and signalling, 4,060 cars, more than 1,200 bridges, major overhaul and depot facilities as well as many other associated structures and earthworks. In addition it manages two generating stations, providing about two-thirds of its own energy requirements and maintains the power distribution system for the whole network."

"But London's advantage of being the first in the Underground business brings with it disadvantages too. Much of the system is aged and, frankly, 'tatty'. A system that had been London's pride and the envy of the world could, over the next decade, become an embarrassment to the nation's capital unless we do more to renew and improve the system. Looking back, and reflecting on the successes and failures of my eight years with the Tube, and there are both, I believe that the failure to get this message sufficiently into the heads of politicians, and other movers and shakers, is the greatest."

One only had to compare some parts of the Underground with the Victoria Line to see what could reasonably be done. Difficult, but essential messages had to be spoken. But before turning to the future it was necessary to look to the recent past, since problems of future uncertainty were more easily handled if we understood where we had come from.

The year 1980 had been one of political and professional trauma for LT. For what they had read in the press they might be forgiven for believing that our services fell apart too. Quite the opposite was true. Train miles operated was up by 3 percent over 1979, and reliability was also up on 1979. These improvements were continuing in 1981. On the demand side however the picture was very different. In 1980, the Underground carried significantly fewer passengers than in 1979 and even more so compared with 1970. Over the years 1978–1980 the Rail business had earned operating surpluses, before provision for depreciation and renewal, of £14 million, £15 million and £28 million as compared with capital expenditure of £47 million, £65 million and £78 million — a creditable performance for most Underground railways around the world — though not, of course, in my previous job in Hong Kong.

Though it would have been good to have more and newer trains immediately, it was important to be renewing District Line stock. We reckoned that 35 years was a good life for our stock. After the 1983 stock that would replace the 1938 tube stock currently operating far beyond its economic life on the Bakerloo Line, we would not be buying new stock until we started to replace the Central Line stock in 1989. The 1990s and early part of the new millennium would require continued heavy expenditure since Northern and Metropolitan Lines trains would be coming to the end of their working lives. So there was a period when we could attack the neglect on the rest of the system. History has shown how important it was to get investment in LU growing again, to match the very rapidly increasing ridership in the 21$^{st}$ century. But the 1980s had indeed been a crucial decade in the turnaround. Lesson — The 1980s was a crucial decade.

Happily we successfully launched a station modernisation programme, made possible by the 'trough' in rolling stock expenditure. However, I did express my concern that if the flak flew across the Thames between Whitehall and County Hall, what would suffer would be investment in the Underground. Public transport played a vital role in the life of the community and was therefore a very live political issue. Its long-term future was too important to be enmeshed in short-term political crossfire.

Of course, national finance was scarce, and the Underground had to take its place alongside hospitals, roads, universities and many other demands. But it was my responsibility to see that the case for the Underground did not go by default. I simply did not believe that maintaining the Underground was low on the list London's or the nation's priorities. The 1980s was a crucial decade. We had lived off the fruits of the wise men of the past. Unless we made a determined effort we would hand a dreadful legacy to future generations. I spent a great deal of time, talking with or giving addresses to politicians, journalists, professionals and countless others about the need for more money.

At the time the annual investment in renewal and improvement was some £80 million. In order to hand on to a future generation the backbone to London's transport network in a fit state, we could and should spend at least 50 percent more than this. The oldest part of the system was nearly 120 years old and about 80 percent was more than 70 years old. Many aged assets remained in service — lifts and escalators (up to 70 years), signalling and cables (up to 50), track (up to 70), and stations (up to 120).

For the first time for several years LT was not extending the Underground network. We simply could not afford to. The two areas to which we were giving most attention were the extension of the Piccadilly line to Heathrow Terminal 4, and possible developments in the Docklands area. The likelihood in the latter case was that rail developments would be of the light, rather than the heavy Underground variety.

So we knew what was necessary, and we were beginning to put a proper process in place. A Railway Strategic Review (LU, 1982) followed and reported that there was a sound basis for planning processes at a tactical level and individual projects and service options were carefully reviewed, but there was no proper process by which major business options were systematically reviewed, nor was any thought given to changed or alternative roles for the business. It was recommended that a rail strategic planning process should be set up on an annual basis. The main phases would be — a rail business planning phase addressing major business options (August–January), a review and recommendation phase contributing to the main board's strategic review (February–April), and working parties and detailed reviews (May–July).

"On 24 July 1981, the MD (Railways) submitted a memorandum to the Rail Board indicating general acceptance of the Booz proposals but advocating that the full implementation of their proposals immediately might involve too much diversion of management time. The Memo confirmed that strategic planning should no longer be conducted within the same framework and timescale as tactical planning, and time would be allocated to enable exploration to be made of real strategic objectives and options."

"The Railway Management Meeting on 15 September agreed that task forces would be set up to study bedrock investment, fare collection, and manpower and employment policy. There would also be specific studies on possible extensions to the rail system, excess capacity in the system and measures to increase demand or reduce that capacity, consequences of an energy crisis in 5–10 years, use of any increased investment funds available in the rail business, and segregation of Northern Line services. The strategic review process for 1982 also took into account the work of various working parties and study groups — mechanical engineering maintenance study, the strategic plan for engineering, the railway investment working party, and the fares simplification working party."

"The studies commenced at a time when a massive fares reduction was imminent and when the external policies — at least for the next four years — seemed likely to involve increasing financial support to keep fares low and services high. The detailed studies were nearly completed when the House of Lord's judgement completely changed the external situation and therefore the framework within which the business has to operate."

"The precise nature of that framework and the consequential financial constraints are still not known but it seems certain that there will be a sharp increase in fares, a reduction in the level of services, and some contraction of the network. The fact that policies can change so dramatically makes strategic planning more difficult and more necessary. It is vital, in a rapidly changing world, that short-term decisions are not only expedient, but also consistent with directions in which the business ought to be going in the longer term. In particular, the work on 'bedrock' investment should be valuable in this respect and the fare collection study was designed to cope with a wide range of charging options."

Thus we had a proper planning process in place at last, but there were a number of specific issues that had to be handled in parallel, involving a number of actions and policy implementations that struggled to get LU moving in the right direction at the same time. In addition, I was still struggling with the mindset of some of my, otherwise excellent, colleagues. An opportunity to 'open eyes' presented itself as a result of my membership of UITP.

### 6.7.1 International Union of Public Transport

International affairs have always fascinated me, increased by studying in the USA. Professional life in the UK allowed me to exercise that interest. My substantive involvement with international organisations included UITP.

It was founded in Brussels in 1885, by 50 representatives of the principal European tramway operators. Its aim was 'the need for engineers and specialists to learn from the experience of colleagues is of the highest interest to all tramway administrations,' which is exactly the spirit of enquiry which needed encouraging in London Underground.

The introduction to UITP, which I regard as the brotherhood (now also sisterhood) of urban public transport, came on joining its Metropolitan Railways Committee from the PTE in 1972. Formally, one was not allowed entry until an undertaking already had a metro. Indeed, Michael Robbins, a predecessor as MD of LU, used to say that the committee was the most exclusive in the world — the entry fee was one metro. Robbins and T C Barker had also written an acclaimed two-volume history of the Underground (Barker and Robbins, 1963 and 1974).

The PTE had been allowed to join because it was already committed to its project and would therefore become an operator in due course. I found this very helpful. As a 'new boy,' it was a privilege to be invited to mix with and learn from Paris, New York, Moscow and others. The Metro Committee was a very supportive organisation. Fierce criticism is part of the 'job-description' for any urban public transport boss, so a regular dose of understanding that one is not alone in the world is very therapeutic. I remained a member until I 'retired' from public transport and became an honorary member. Lesson — Find support by learning from colleagues around the world.

In 1982, a Finance and Commerce sub-committee was formed and I became its chairman. I have always found it valuable to make comparisons with others. It certainly teaches important lessons, but it also provides perspective. One of my favourite memories was of a presentation I made to the UITP in which, in order to make a point about the size of the Underground, I superimposed a plan of it on top of a map of Holland. It connected all of the major Dutch cities. A colleague, who then managed the Brussels system, said to me afterwards, "Tony, if you overlaid the Underground on Belgium, it would stretch from Holland through Belgium and into France!" That story did more to illustrate the challenge of the Underground than any other I know.

We were, of course, rather more scientific. At the 1983 Congress of UITP in Rio de Janeiro I presented, together with Hans Meyer of Hamburg, the sub-committees' paper on measures of productivity in 26 undertakings (Ridley T *et al.*, 1983). We opened the Introduction by saying that, 'previous reports from the metro committee to UITP Congresses have concentrated discussion on technical issues and planning strategies. In this report, for the first time, the economic use of resources as measured by the productivity of underground systems is to be considered.' Apart from the results themselves, it was interesting that only two of the 26 undertakings, those of the two authors — London and Hamburg — were willing to be named in the tables of performance measures. All the others remained numbered rather than named.'

A particular value of the work to me was to encourage a more searching examination of our own performance in the Underground. No amount of special pleading about age, size, tunnel diameter or corridor length could hide the fact that, on a number of measures, Hamburg outperformed London. A fraternal visit to Hamburg followed and the lessons learned were built into the Underground strategic planning process. This focused on a whole series of initiatives that involved changes in the company culture.

When the Hamburg report was published I handed copies to all of my senior colleagues in the Underground. They had had a multitude of reasons as to why Hamburg performed better than London on nearly every measure. Of course, there were some special circumstances. My personal experience made it very easy to understand why London, the oldest metro in the world, could not hope to outperform Hong Kong, one of the newest. The team that went to Hamburg came back with the scales lifted from their eyes. Not that this implied that LU could be miraculously improved, but minds had been opened to possibilities — a splendid example of learning from colleagues around the world.

A particular benefit of membership of the UITP Metro Committee had been to travel, at six monthly intervals, to meetings hosted by the management of metros in the important world cities, which took me to Oslo, Sao Paolo, Washington DC/Atlanta, Lisbon, Barcelona, Montreal and New York — an important learning experience in each case. Happily, I was able to keep these links when I created the RTSC at Imperial, with its metro benchmarking clubs.

At about the same time I became President of the Railway Study Association and named my presidential address, 'Urban Transport — Cost Reduction through Investment' (Ridley T, 1983). Throughout my career I have stayed in touch with various professional and academic bodies, which gave me regular opportunities to sum up my thinking to date in front of friendly but questioning audiences. I commend the practice. It excellently focuses the mind. I said,

"When I reflect on railways in London there is one fact which seems to me to dwarf all others, that each working day between 7am and 10am, of more than million journeys, no less than 700,000 passengers arrive in Central London by rail with, noting that there is some double counting where passengers use both rail modes, about 450,000 by Tube, 400,000 by BR, with roundly 100,000 by bus and 200,000 by private transport. It seems to me that all other facts pale, almost, into insignificance compared with that 700,000 and it also seems patently self-evident that, for the rest of our lives and at least until well into the next century, conventional railways will be required to carry the major proportion of the vast number of commuters who will travel into Central London every day. Greater London, the southeast and indeed the nation depend on Central London and London depends on the existing rail network as the backbone of the total transport system."

### 6.7.2 Tube Costs

In 1980 the Tube offered 30 million train miles in service, the same as in 1970; place miles were 24.4 billion, very nearly the same as in 1970 (24.6 billion). But passenger journeys had fallen to 559 million (from 672 million in 1970), and were to fall further to 498 million in 1982. Over these years costs had leapt. Staff costs had increased, 1970–1980, from £32.6 million to £181.9 million (or £116.8 million at 1970 prices), an increase 50 percent faster than inflation (Ridley T, 1982). The greatest increases had taken place in 1974 and 1975. In those years LT launched a special recruitment drive, together with improvements in pay and conditions, to counteract a severe staff shortage at the time. In addition, there had been a gradual improvement in leave and superannuation provision, together with increased payments required to be paid by LT in respect of National Insurance. The evidence suggested that the extent to which staff costs had run away beyond the rate of inflation was not because of increases in basic rates of pay, but rather other 'improvements'. Thus, although various productivity improvements had been made, the business had in effect had to run in order to stand still.

Because of the labour intensiveness of public transport, the need constantly to strive for improved productivity and effectiveness was self-evident. Even if LT were to double the amount of subsidy that it could attract, there would still be a need to tackle the cost problem.

## 6.8 Major Projects

We already knew that we had a major rolling stock renewal programme ahead, but not until 1990–2000. We also had to renew equipment for control and communications, signalling, track and structures and depots but, because of the 1980s 'trough' in rolling-stock orders, we were able to press on with a large programme of station modernisation. Stations had been sadly neglected over the years and, not only did we make stations brighter and more attractive, they were also easier to keep clean and smart. A number of different developments came together on the Central line that was due to be renewed in the 1990s. Also decisions were necessary for the future of Acton Works, and for power generation (then at Lots Road and Greenwich) and supply.

In all of these our programme was divided into 'bedrock' (that investment that was essential to prevent the deterioration of the system) and 'other' (that investment that would improve the performance of the system for the passengers, through better reliability, higher speed, improved communication, and more attractive stations). We coined the term 'bedrock' investment to describe the absolute minimum required to keep the system going. Then there were a whole series of investments necessary to make the system presentable, and attractive, to the passengers. But this represented 'business as usual,' or perhaps somewhat better than usual. This was, however, was not good enough. Although all of those investments were necessary, indeed essential, for the well-being of London, they did not of themselves attack the cost problem.

### 6.8.1 Strategic Initiatives

We had identified six strategic initiatives that would have the most important impact — One Person Operation, Enhanced OPO, Underground Ticketing System, Station Staffing, Engineering Productivity and Marketing. In the engineering departments, we went through all of our activities to see where we could reduce costs and every aspect of maintenance, design and administration was examined by a Productivity Task Force. Savings were anticipated from changes in staffing levels but, in engineering departments, they were offset by an increase in the capital renewal and improvement programme and the need to maintain new, more sophisticated equipment.

In operations, major savings were possible, also dependent on investment. A radical improvement in the UTS was planned as were savings from changes in station staffing levels. The UTS project was particularly successful — a combination of new technology together with building works at a large proportion of the stations on the Underground. While not as glamorous as a major civil engineering project it was both successful in its own right and paved the way for future ticketing developments. I was fortunate to have Tony Humphrey as an effective project manager, a member of the Hong Kong team who had returned to LU.

We also reviewed train operations. Significant savings in costs could be achieved by reducing the number of staff traditionally employed to operate a train — the driver and guard. The first move in this direction on LU was when the Hainault–Woodford branch of the Central Line was converted to experimental Automatic Train Operation prior to the introduction of a similar system on the new Victoria Line.

Planning for the sub-surface lines proceeded on the basis of OPO. Thinking for the tube lines had been for some form of ATO but, because of the high cost of converting existing signalling, this did not prove to be cost-effective. Attention therefore focussed on enhancing conventional OPO in tube tunnels. Initial discussions with the Railway Inspectorate were encouraging and it was proposed to develop the system (enhanced OPO), which was highly cost-effective, for implementation on those tube lines — Bakerloo, Jubilee and Piccadilly — equipped with conventional signalling and operated by rolling-stock capable of being readily adapted for OPO.

ATO, although considerably more expensive than enhanced OPO, did have a number of advantages and its cost was likely to reduce significantly as a new form of track circuit, capable of handling coded command information, was developed and became an integral part of any signalling modernisation. Initial development work on an up-to-date ATO system was underway with a view to carrying out a detailed evaluation of the costs and benefits as soon as possible. If justified, it could then be adopted for use on the two remaining tube lines — Central and Northern — which would be the first to be equipped with modernised signalling incorporating the new track circuits.

These were the directions in which it was believed that the Underground had to move if it was to become, not only modern and attractive to the customer, but cost-efficient as well. Although I believed that a higher level of subsidy was justifiable, more in line with so many other cities of the world, it would have to go towards keeping fares down, rather than underpinning higher costs. We planned to carry forward 'bedrock' investment, system improvements, passenger improvements and strategic initiatives in parallel, and it was essential to do so. An increasingly efficient but ever more tatty Underground was no more acceptable than a bright and attractive Underground whose costs were out of control.

Meanwhile, we were also developing the relationship between BR and LT. From the time that the two Sir Peters, Parker and Masefield, Chairmen of BR and LT respectively, created a small piece of history by appearing together in February 1982 before the House of Commons Select Committee on Transport, the relationship was excellent. BR and LT had worked together before, but there was now a new philosophy that, in spite of complex institutional arrangements between us and the financial climate, we pressed on as professionals working to create a single transport system for London. Some of it was provided by BR and some by LT. In the ideal situation the passenger would barely notice the difference — simply that there was a clean, modern, efficient and cheap, integrated public transport system. To achieve that we were not developing a vast new railway plan, but seeking to put to better effect what we had already. The extension of through-ticketing and the development of the London Connection brand were small examples. But, following the successful introduction of LT's Travel card in May 1983, we were on the verge of the first stage of an integrated fares structure for BR and LT. While much detail remained to be worked out, we knew that we were within reach of something London should have had years previously. Throughout my time with LU I had the good fortune to have as opposite numbers in BR three senior colleagues and friends — Geoff Myers, David Kirby, and most lively of all, Chris Green.

## 6.9 Organisational Change

In 1984, the Tories passed the London Regional Transport Act creating LRT, which came into being on 29 June 1984, reporting to the Secretary of State for Transport, my namesake Nicholas Ridley (no relation). Some said that this was the only 'nationalisation' pushed through by Thatcher, the arch privatiser. In 1985, London Underground Ltd. was set up as a wholly owned subsidiary of LRT and I became the executive Chairman and MD. This was one of the serious mistakes of my career. Three years later, after King's Cross, I appointed Denis (now Lord) Tunnicliffe as MD, while I continued as Executive Chairman. The appointment of a separate MD is something I should have done when LUL was formed. By then I was no longer 'managing decline' but looking ahead to the challenge of ever-increasing ridership, addressing what major investment was required and how it might be financed. In addition, I was personally driving the DLR project forward. It was all too much for one person.

### 6.9.1 Smoking on the Underground

One of the issues that lay heavily on my mind at that time was the question of smoking on the Underground. It should be borne in mind that this was long before the acceptance today of banning smoking in very many public places. During the time of GLC control we had developed a proposal for the introduction, on an experimental basis, of a ban on smoking on trains. This had caused a substantial row and we were castigated in the press, and directly by letter, by Forest (the Freedom Organisation for the Right to Enjoy Smoking Tobacco), although ASH (Action on Smoking and Health) took the opposite side of the argument. The ban was due to be introduced in July, almost immediately after the creation of LRT.

Nicholas Ridley was known to be an avid smoker, and was a friend of Ralph Harris (later Lord Harris of High Cross) of the Institute of Economic Affairs, who became Chairman of Forest in 1987. Nicholas, when he obtained the powers, quickly swept away Ken's Board appointments and replaced them with eight of his own.

On the evening of 23 November 1984, a fire broke out at Oxford Circus station, in which 14 people were taken to hospital and were treated for 'smoke inhalation'. Of these four were passengers, one a policewoman, and the other nine members of staff. With one exception, all were released, after treatment, that night or the following day.

In October, LUL issued a statement on the circumstances surrounding, and the actions following the fire. Some 50 recommendations had been made as a result of the fire and, by the date of publication, 27 had been completed, the remainder all being progressed. One particular recommendation made by the London Fire Brigade was that there should be a complete ban on smoking on Underground premises. In its wisdom, or not, the new Board of LRT had not accepted this recommendation but did confirm the continuation of the experimental ban, and extended it to all below-ground stations.

In the 9 August 1984 edition of NCE, I had been interviewed under the heading, 'Shake-up for LU.'

"Ridley does not think the coming of LRT will have any dramatic effect on the way the Underground operates. "What I expect in future are things that were already being planned under the LT banner. There will be changes in emphasis rather than drastic alterations in the way we work. When we were under the control of the GLC we were given tremendous support to develop the Underground. But the Council's main interest was maintaining employment while becoming more efficient. It is wrong to pretend that one end of the political spectrum believes in efficiency and the other one doesn't. But the emphasis is different." A strategic plan for the Underground's next 20 years drawn up under LT still holds.

'For reasons of safety great emphasis is put on track renewal. Ridley draws attention to the high frequency of derailments in New York as an example of the dangers that can occur if this area of work is neglected. One of the things that distressed everyone on the Underground was that track was allowed to deteriorate during cutbacks by past GLC administrations. London's tracks did not get to the point of being unsafe, but unless track is renewed constantly real problems are stored up.' Bridge replacement is another substantial area of work for LRT civil engineers and Ridley insists that station modernisation should not be neglected if huge structural maintenance costs are to be avoided. Relocation to this growing area of work on station modernisation is being offered to many civil engineers currently employed on the almost completed terminal four (Piccadilly) tube extension at Heathrow. "Station modernisation is less glamorous than tunnelling in clay but involves difficult and intricate replacement work such as pushing through escalator replacements in very confined spaces. This is probably more demanding than new construction."

At the same time a presentation on strategy was made to the Board of the new LRT (Ridley T, 1984). The situation that I inherited on becoming MD, I said, was a number of adverse long-term trends — declining population and employment, rising real fares levels, rising real costs and falling ridership. I began,

"The Underground is a complex system which has developed over the last 121 years into one of the largest mass transit networks in the world. This presentation has been designed to raise some of the important issues that face the Underground today. Special emphasis will be paid to where the Underground is going in the future — a future that we believe is a bright one. A number of problems will have to be addressed in the next five years and beyond — not least all of the management of change and industrial relations. LU has developed a strategy for dealing with the next five years up to 1990 and this forms the basis of the presentation."

### 6.9.2 Planning for Renewal and Growth

Starting in 1982 we had drawn up plans to combat these trends, plans that represented a radical departure from the past reactive approach. We had reviewed the strategic issues confronting LU and those that could be foreseen over the 20-year future. We examined a number of alternative courses of action. It was clear that LU could not choose to 'do nothing'. Remedial action was required to overcome rising costs and falling passenger receipts.

Thus, the 1984 Strategic Plan called for more production at less cost; with fewer staff; *and* more quickly than was previously proposed, with targets by 1987/1988 including — a reduction in unit costs of 6 percent; the maintenance of the high quality of the product with 98 percent of scheduled miles operated; and an increase of 100,000 train miles for extra services to Terminal 4 at Heathrow, and increases of the service to Harrow and Wealdstone on the Bakerloo Line.

The proposed UTS would provide a completely new ticket issuing system covering every Underground station. In the central zone, ticket control would be by means of automatic gates. In the suburbs control was to be based on open stations and increased inspection on trains, backed up by penalty fares. The system was designed so that automatic gates could be installed at further stations when cost comparisons with open systems justified them.

The Engineering Productivity Task Force had been set up to carry out a root and branch examination of engineering processes, to identify and then implement potential productivity improvements. One in particular involved the transfer of train overhaul from Acton Works to Line depots. The evolution in the design of modern rolling stock had both reduced the requirement for routine maintenance and increased the intervals at which train overhaul needed to be carried out. Consequently, Acton had become too large and ill-adapted for its purpose.

A Study Group had been set up to determine whether maintenance and overhaul procedures could be altered so as to avoid the need to spend substantial sums of money in restoring, adapting and bringing Acton Works up to modern standards. Among their recommendations the principle one was that train overhaul and wheel-set profiling should be transferred to depots; also that a new workshop should be set up for competitive work areas. LU had accepted the recommendations subject to further discussions with the trade unions.

They had voiced strong opposition to all the proposals but, in particular, that relating to transfer of train overhaul to depots. A Joint Working Party (management, trade union officials and local staff representatives) had been set up to examine the way in which Acton might be made more viable so that work might justifiably be retained there. Meanwhile a timetable for implementation of LU's final plans was temporarily in abeyance, but it was thought that it might be possible to commence transfer of train overhaul to depots in 1985 with the commissioning of a new workshop in 1988.

The presentation concluded that the Strategic Plan was primarily a blueprint for change. The difficulties should not be underestimated, but it was achievable, and the future well-being of the Underground depended on its achievement.

### 6.9.3 One Person Operation

Throughout all of our planning the railway had to run every day and continually improve its performance. There were two major challenges — OPO and Acton.

OPO was clearly an integral part of improving productivity. What amazed me, when I joined LU, was to discover that there had been an agreement between management and the trade unions for some ten years, but it had never been implemented — just left in the 'too difficult box'. What was worse, it had implications elsewhere in the organisation. When I started to talk to Leslie Lawrence, the Engineering Director, about productivity in his departments he said, in effect, 'I promise you one thing. Until you implement OPO you will never get any improvements in engineering'. What he meant was that there was a massive psychological and managerial block about the determination, or lack of it, to make beneficial changes — in other words, the introduction of OPO was the key to changing mindsets elsewhere in LU.

In fact, by 1985, we had made more progress than anticipated, so much so that we were ready to introduce OPO on the East London Line, one year earlier than anticipated, with a number of lines having already been converted. As was usual in the world of industrial relations, this did not mean that the unions would readily accept proposals in individual cases. Thus, as we moved towards the due date of Monday, 20 May, strike action was threatened, but it did not seem to be sensible to me for the unions to strike when so few trains and staff were affected, and when proposals had already been implemented elsewhere.

### 6.9.4 Involvement of the Trade Unions

During the 1980s, there were three principle Trade Unions with whom I negotiated — the National Union of Railwaymen (NUR), General Secretary Jimmy Knapp; ASLEF, General Secretary Ray Buckton; and the white-collar TSSA, General Secretary Tom Jenkins. The normal practice was for the General Secretaries to handle negotiations with BR, delegating business with LU to their deputies. Nonetheless, I had dealings with each of the three on particularly important or contentious issues.

Jimmy would frequently bluster, and although I liked him when I could understand his nearly impenetrable Scottish accent, he was not a strong leader. Negotiating with 'hard men' who can deliver is far better than negotiating with those who cannot. "Oh Tony", he would say, "don't tell us you've not got enough money. Get across to the Department of Transport and get some more". Or, "we know you get your orders from Mrs. T through your cousin Nicholas (Ridley, who was the Minister)". I was not related. Nicholas was the younger brother of Lord Matthew Ridley, of Blagdon Hall in

Northumberland, who had been Chairman of Northumberland County Council and Chairman of the Steering Group for the T&W Transportation Study. Matthew was subsequently to confer an honorary doctorate on me at Newcastle University, in his capacity as Chancellor. "Well done the Ridleys," he said.

I would have preferred to delegate all negotiations, but the Unions always demanded to deal with the 'top man', particularly in the annual wage negotiations. This was always the least attractive of a very exciting, but demanding job. I began to feel that Industrial Relations were a 'conspiracy' between management and the unions, against the public. Wage negotiations had a limited timescale, but everything else seemed to have no end point, and could drag on and on and on.

Bizarrely, one dispute introduced me, briefly, to Robert Maxwell. We had made good progress with the implementation of OPO but, when it came to the East London Line, Jimmy decided to call a strike. The new arrangement was due to start on a Monday and I had arranged a final meeting on Sunday morning in a last attempt to avert it. I was at home on Saturday, mowing the lawn, when Jane called me to the phone. "It's Robert Maxwell", she said.

Maxwell introduced himself, in his very plummy accent that hid any vestige of his Czech background, and said that he had learned of our problem. He said he would like to help. "I can arrange a place where you can meet Knapp, absolutely confidentially." I thanked him and said that I already had a place and time to meet Jimmy. He pressed his case. "I don't want to teach my grandmother to suck eggs, Dr Ridley, but …" When I declined the offer of assistance he said that we must have lunch together when the problem had gone away. I never heard from him again.

My Sunday meeting did not resolve the issue and, on Monday morning 20 May, I was in the control room at 5.30 am, counting trains coming out on to the system. When by 10 am we had a reduced but satisfactory service I knew it was all over. At 2 pm I got a call from Jimmy, "Tony, I'm going to do the most difficult thing I've ever had to do, call off the strike." That's alright, Jimmy," I said. "I won't be triumphalist in the press."

Interestingly, strikes in Newcastle were of no consequence to the government. But, when there were threats in London, the civil servants would be on the phone at frequent intervals, "Of course, I don't want to interfere with your responsibilities, but I must keep my Minister informed."

The press generally panned the NUR. *The Times* (21 May 1985) described the collapse of strike and said that Union leaders had called off the strike, called in defiance of a High Court ban, because fewer than 6,000 of the union's 15,000 members operating Tube services had obeyed the strike instruction. They quoted me as saying that we were pleased for both our passengers and for our own staff, who had demonstrated that they wanted a secure future in an increasingly expanding system. *The Guardian* (21 May 1985) wrote that LU was able to run a 70 percent service on early rush-hour trains, and on the previous evening the service was running at more than 80 percent. In their editorial they pointed out that, on the East London Line, Union members crossed picket lines to work the new system that the (NUR) executive found intolerable.

### 6.9.5 New Workshops at Acton Depot

In the year between the King's Cross fire, November 1987, and my resignation from LT in November 1988, much of my time was taken up with ensuring that morale held up and that we continued to provide good levels of service. In addition, the Inquiry was heavily time-consuming. But there had been another crucially important task facing the Underground. In 1980, ridership had been falling steadily for more than 30 years — from 720 million journeys p.a. in 1948 to 498 million p.a. in 1982. Then, over a 5-year period, it rose by no less than 60 percent to 800 million p.a. The task that I inherited in 1980, of rescuing a declining system, had been transformed into the new challenge of coping with massive overcrowding. It became necessary to repeat the strategy studies of the earlier half of the decade, but under completely changed circumstances.

OPO introduction was one of our landmark actions in moving towards the modernisation of the Underground and, indeed, of the LU organisation. So equally were the changes we made at Acton. In June 1986 we received a letter from the Department of Transport, (DoT to LRT, 25 June 1986).

"I am writing to let you know that Ministers have approved the expenditure of £15.22 million (at December 1985 prices) on the provision of a new workshop at Acton, for which authority was sought in the submission on 11 March 1986 and subsequently discussed extensively with the Department. In approving the project, Ministers were concerned that the competitive position of Acton, as against the private sector, should be kept under review. We will want to discuss what mechanisms you are proposing for this, bearing in mind the continuing duty under section 6 of the 1984 Act to invite tenders for work, where appropriate; and we will be looking for an assurance that in-house costs will be regularly tested against outside competitors."

Our press release (LUL June 1986) said,

"A new £15 million workshop is to be built by London Underground at Acton. It will overhaul equipment from LU trains and is scheduled to begin operations in 1989. Modern train design has greatly reduced the amount of maintenance required. This meant that the existing 60-year-old Acton Works, which until recently overhauled both train equipment and the trains themselves, had become uneconomic. A detailed study in 1983 showed that train overhaul could be carried out at much lower cost by the depots on each of the Underground lines. Following a successful trial at Golders Green, this transfer of work is under way. The study showed that while some equipment overhaul work could be performed more effectively by external contractors, Acton Works was more efficient than outside industry in some areas and, with new working practices, would remain competitive with industry for many activities. A detailed agreement on new efficient working practices has been negotiated with all the trades unions concerned, making possible the investment in the new factory. The government has approved this investment on the basis of proven competitiveness with industry and LU will regularly monitor production costs against those of industry. Preparatory work begins in August, when the present trimming shop and other activities will be re-located to enable construction to begin in November."

It was important to meet those people who had been most involved in the planning and negotiations — staff representatives of the Unions (EETPU, TASS, AEU, T&GWU, NUR), and of the supervisors and administrative staff; together with management representatives led by Ian Arthurton, Mechanical Engineer (Maintenance), and later Operations Director of LU. I started by saying, "We've come a long way from the doll's house", this being a reference to inefficiencies uncovered by the *Evening Standard* at the beginning of the decade, when staff were found sleeping, playing cards and, in one case, building a doll's house. I went on to thank everyone for the important part they had played but emphasised that the challenge now was to 'make it all work'.

Writing to the Minister for State for Transport, I quote (Ridley T to D Mitchell, 30 June 1986),

> "Dear David, We were delighted to have the news of the government's blessing of our proposal for investment in the Equipment Overhaul Workshop at Acton. I thought I should let you know that I was at Acton on Friday last to talk to our managers, trade union officials and shop stewards. They were, of course, relieved. I think you would have been encouraged by the enthusiasm with which they view the future. Many of the people with whom I consult and negotiate have been under attack from several of their colleagues — attacks sometimes of a political nature — particularly because our proposals have involved major changes in work practices and substantial reductions in employment. Acton has been a significant chapter. As I told you when we last met. Its lessons are already, very positively, being transferred into other parts of the Underground with beneficial results."

## 6.10  1988 Strategic Plan

In 1986, Booz Allen were appointed to carry out studies that would allow us to develop a new strategy for the future, the 1988 Strategic Plan. They were supported by a small LU team including Hugh Sumner, of whom more in Chapter 8, to work with them. It was absolutely necessary to produce a new strategy in its own right, but in 1988 it had the additional benefit of raising the sights of the management beyond the traumas post-King's Cross.

A presentation at a special LUL board meeting in February 1987 listed the strategic issues that were to be faced, most especially because of the extraordinary increase in ridership that had taken place over five years. These included:

- The system was not operating at optimum levels in terms of capacity, quality, safety, cost efficiency, organisational efficiency and financial performance;

- There was limited ability to develop market share and revenue;
- There was inadequate forward planning;
- The existing network could not cope with projected growth, even if the network were operated optimally;
- Safety, quality, and reliability would get worse unless access was restricted;
- Major fare increases to constrain demand were probably unacceptable;
- Consequently, the development of London was likely to be constrained by inadequate transport infrastructure.

LUL had limited potential to meet projected demand within its own resources because solutions were expensive and technically complex. Current financial losses were likely to continue to persist due to safety and quality problems. It had no control over its fares. Debt financing was prevented by its existing legal framework. All of this was compounded by uncertainties over possible future legislative changes.

These issues had to be considered in light of LUL's company mission statement, requiring it to be efficient, well managed, and customer-oriented operating within a clearly defined framework; to provide a safe, modern, good quality rapid transit network and service; properly planned to meet changing market conditions; and contributing to London's long-term future prosperity.

### 6.10.1 Impact of the King's Cross Fire on the Process

The issues had to be faced and everyone was willing, but the task was made more complex when, immediately after a second special Board Meeting on 16 November 1987, we had to focus on the aftermath of the King's Cross fire. I was determined that, although my priority was the follow-up to King's Cross, work would proceed and we held a further special LUL Board Meeting on 7 January 1988. Then, after the conclusion of the Inquiry programme, we were able to focus on the 1988 strategic plan in June of that year.

The work had been set in motion prior to the King's Cross fire but, obviously, it was wise to revisit the original objectives and management challenges, in the light of the fire. The Board meeting in January 1988 high-lighted and re-emphasised the importance of sound asset management — accurate records, performance specifications, contract control and analysis of asset performance. The increased ridership was putting greater pressure on the increasingly aged infrastructure. Thus, maintenance, renewal and upgrading were assuming yet greater importance than they had when fewer passengers were being carried. There was an urgent need to find solutions to increasing system congestion. The work had to be managed in an ever-busier environment. Various plans, both short- and long-term, needed to be revised.

In the midst of all of this, and given the possible loss of confidence of the public in the organisation, it was imperative to re-energise a customer-focussed approach, putting increased demands on managers and staff. A programme of asset condition surveys was organised as was a zero-based definition of the core organisation. To tackle the capacity shortfall, three options were analysed — 'do minimum', extreme retrenchment, and extreme expansion. The business, by now, was looking ahead to the time when LU would be carrying one billion passengers per annum — an unbelievable figure from the perspective of 1980, when I returned from Hong Kong. Yet, by 2010, LU had carried 1.1 billion passengers in the year, which has increased further since.

Meanwhile it was important to maintain and develop contact with the press. I also pursued a media relations programme, organised for me by Michael McAvoy of McAvoy-Wreford-Bayley, from which it became clear that newspaper editors were available for lunch at the Savoy. Over six months from May to November, I met, and briefed more than 20 editors and other senior journalists, including Wilson from *the Times*, Owen of the *Financial Times*, Pennant-Rea of *the Economist*, Burnett of *Independent Television Network*, Preston of the *Guardian*, Whittam Smith of *the Independent*, Wyatt of *BBC TV*, and Greg Dyke of *London Weekend Television*.

Charles Wilson brought Rodney Cowton, his transport correspondent, with him. He wrote two pieces under the headings, 'Underground Chief Draws up £3 billion Plan for London', and 'Underground Steps up Search for Private Cash'. (*Times*, 16 and 17 May 1988) It was only with Peter Preston that the conversation dwelt at all on the fire. I already knew Peter from the past, when we both lived in Dulwich during the latter half of the 1960s. To his credit, he was the only journalist who told me that the Inquiry could well end with my resignation.

I had presented a memorandum to the LRT Board, LUL 1988 Strategic Plan (Ridley T, 1988a), which addressed our thinking to that date. It said

> "The Company has undertaken a fundamental examination of its business and options for its future objectives and strategy. Four major options have been developed which result in different financial performance and transport benefits over a ten to fifteen year timeframe. LUL have developed a recommended strategy based on the results and insights from four cases and an overview of past trends and related company performance. The aim is to establish a 'New Contract' with LRT and the Government, through the Department of Transport and the Treasury, which will formalise the Company objectives and targets to be achieved by implementation of the agreed Strategic Plan."
>
> "The purpose of this memorandum is to seek the Board's agreement to the objectives and strategy proposed, and to start the process of reaching agreement with the Government."

The 1988 Strategic Plan was produced in June. Its introduction said that the Underground was a vital part of London's past and future, and LUL provided the extensive rapid transit infrastructure and the high-density passenger rail services on which the economic success of London depended. LU's performance was significantly influenced by the performance of London and the Southeast. Three external factors had been dominant in the past fortunes of LU — economic activity growth and inflation, demographic and social changes, overall government policies towards public transport.

These external influences were expected to continue, and LU's response would determine its overall performance. Conversely, the way in which LU provided transport services in the future would help to determine the ability of London to maintain and improve its position as one of the world's leading trading/financial centres.

Over the previous 20 years, real wealth had increased, markedly in the last 5 years; population had declined by 15 percent and employment by 23 percent, but both were starting to grow; heavy industry had closed or relocated, while light/service industry had grown; population and employment patterns had changed significantly; commuting to and from London had increased; rising car ownership had increased road congestion; increased rail commuting had led to congestion in the centre of the network and on key radial routes.

In that time LU's performance had mirrored these external changes. Fares levels had outstripped inflation by 1.5 percent, but had fallen below average earnings by 0.4 percent. Unit costs had risen significantly during the 1970s, despite relatively low pay levels compared with those for competing employment opportunities. Investment priorities had varied between network expansion in the 1960s and infrastructure renewal in the 1980s.

As a capital intensive business, LU had limited ability to respond to rapid shifts in market demand, resulting from demographic characteristics and changing peak demand. It also had a limited ability to obtain a reasonable rate of return on the large asset base. Labour practices, attitudes and agreements were outdated and there was a general attitude of resistance to change.

The strategy proposed in June 1988 was a departure from previous practice. It would require change within LU at all levels and require new relationships between LU, LRT, government and the travelling public. The strategy recognised that the financial pressures on LU were real and not likely to diminish in the future. It was based on improving LU services, assets and infrastructure to serve a growing customer base and help to maintain the vitality and economic well-being of London.

There were three overlapping principal phases. Phase 1 would focus on internal management reform, service quality, safety improvement, securing sound financial performance and clarification of the external framework within which LUL would operate. Phase 2 would consolidate the changes in Phase 1 and target further quality and reliability improvements across the whole of the existing network, aiming to complete the process of restoration of the system begun in the early 1980s. Planning would focus on preparation for Phase 3, which would concentrate on expansion of the network through major investment in new lines aimed at providing extra capacity in congested corridors. Each Phase was estimated to last some four years.

My final involvement with LU strategic development came on 19 October, when making a presentation (Ridley T *et al.*, 1988) to Michael Portillo, then Minister of State for Transport, when I told him that LU had recovered from the decline of the late 1960s, 1970s and early 1980s. Ridership was up by 60 percent and at record levels. The fares structure had been simplified, with popular new ticket types introduced and fare levels held constant since 1983. Revenue was up 16 percent in real terms, while efficiency had improved by 15 percent. Investment and re-investment had doubled and was now at a record £245 million p.a. — after several decades of under-investment, with increasingly aged infrastructure to keep updated, and with a dramatic improvement in our ability to deliver major projects on time and within budget, witness the Underground Ticketing System.

This unprecedented reversal had brought with it major problems of success and demanded a fundamental examination of the strategic issues, which LU had started in early 1987. LU now faced five strategic issues.

- Much of the network was capacity constrained during peak periods.
- Capacity constraints, and resulting congestion, were causing service quality problems, resulting in safety problems if nothing was done.
- Growth in demand was forecast to continue — some 20 percent in the peak and 30 percent in the off-peak by 1997/1998.
- Major capacity additions required large investments and long lead times.
- LU had neither the cash nor the financial powers to fund extensive new investments in the short or medium term.

I explained to Portillo how we had developed a way ahead, in three phases to be implemented over more than a decade. Our key conclusions highlighted the difficult trade-offs that would have to be made across the business. High fare increases would reduce congestion and allow higher service quality, and produce higher revenues to fund investment. Service volume, quality and safety would be adversely affected if demand growth continued unchecked and capacity constraint was not addressed, and if investment levels were not related to the volume of business. Investment requirements were directly related to fare levels and latent demand, and volume, quality and safety of the service provided. Financial performance was dependent on all these factors which must be considered together, since it was impossible to 'have everything'. Cost reduction objectives had to be consistent with objectives for quality, safety and volume of business.

Portillo's reaction was to say that, following the events of the last year, LU now needed certainty about where it was going, and the Plan made a major contribution to this. He expressed his intention to have objectives agreed for LUL by the end of the year, and he agreed that a flexible approach was essential to ensure that realistic targets were set quickly. No doubt he already knew what was in store for me, and could anticipate my departure three weeks later.

For myself, I consider it an achievement that, in the midst of the aftermath of King's Cross, I was able to leave behind me a sound way ahead for London Underground. It was a very well-considered Plan which, no doubt, required careful examination and, perhaps, modification and also posed political difficulties. But none of us could have foreseen the sad fact that a future government would believe it is necessary to impose the ill-conceived and disastrous PPP on the Underground.

## Bibliography

Allen B (1981a), Development of a strategic planning process for the London Transport rail business (unpublished), Progress Report, 30 April.

Allen B (1981b), Development of a strategic planning process for the London Transport rail business (unpublished), Draft Final Report, June.

Barker T C and M Robbins (1963 and 1974), *A History of London Transport*, Vol. 1 — the 19th century, Vol. 2 — the 20th century to 1970, George Allen & Unwin, Crows Nest, Australia.

Carvel J (1984), *Citizen Ken*, Chatto & Windus, London.

DoT to LRT (25 June 1986).

Forester A *et al.* (1985), *Beyond our Ken — a guide to the battle for London*, Fourth Estate, London.

LTE (1978), Reorganisation of LT 1978–1979 (7 Dec).

LTE staff notice (25 July 1980).

LU (1981), Strategic plan for the railway business (unpublished), March.

LU (1982), Railway strategic review (unpublished), January.

LUL press release (June 1986).

NCE (1980), Ridley takes the Tube, 20 November.

NCE (1984), Shake-up for LU, 9 August.

Ridley T (1981), The future of the Underground, in *Centenary Briefing*, Commerce International, London Chamber of Commerce, pp. 38–39.

Ridley T (1982), Counting the cost of public transport, Annual Conference, Institute of Transport Administration, Manchester, 26 October.

Ridley T (1983), Urban transport — cost reduction through investment, Presidential Address, Railway Study Association.

Ridley T (1984), LU strategic plan (unpublished), Presentation to LRT Board Members, 21 August.

Ridley T (1988a), LUL 1988 Strategic plan: Summary, Memorandum for the LRT Board, 29 April.

Ridley T (1995), Customer focus — the international dimension, in *Railway Study Association Bulletin* No. 47, pp. 14–32.

Ridley T *et al.* (1983), Productivity comparisons between metropolitan railways, 45th International Congress, Rio de Janeiro, UITP, Brussels.

Ridley T *et al.* (1988), 1988 Strategic Plan (unpublished), Presentation to Michael Portillo, Minister of State for Transport, 19 October.


## Press

*Daily Mail* (18 June 1980).
*Evening News* (17 June 1980).
*Evening Standard* (17 June 1980).
*Evening Standard* (20 June 2000).
*Guardian* (21 May 1985).
*Times* (21 May 1985).
*Times* (16 and 17 May 1988).


## Personal Correspondence

Cope C to T Ridley (20 February 1979).
Mote H to T Ridley (23 February 1979).
Ridley T to D Mitchell (30 June 1986).

# 7

# Management after a Disaster 1987–1988

What is it like to live through the trauma of a major disaster and its aftermath? What are the priorities after such a disaster?

The fire at King's Cross Station and its aftermath, including the Public Inquiry and its Report, have been included as a separate chapter. By its very nature such an event presents very different challenges and lessons to learn from such a tragic experience — not least that the Underground had to continue to serve the public throughout.

Some lessons

- Management is about maintaining the highest rate of change that the organisation, and the people in it, can stand.
- Looking back, it was unwise ever to hold the posts of both Chairman and MD of LU at the same time.
- Seek professional Counselling if you feel it is necessary and accept support when given.
- Ideally, public Inquiries about accidents should not begin until any scientific investigation, for causes, has been completed and reported.

## 7.1 Introduction

On 28 February 1975, a crowded LU train drove into a dead-end tunnel at 40 mph at Moorgate station (Holloway, 1988). More than 40 people died. This was world-wide news but, being on the other side of the world where I had just started as MD in Hong Kong immersed in a challenging new job, my mind was not on the detailed coverage in the British papers of the tragic accident and its aftermath. Five and a half years later, on becoming the MD of the Underground, the organisation had still not fully recovered from the corporate trauma that that accident caused.

Family experience, had taught me that engineers and operators had constantly to be aware of the potential safety problems of operating equipment. My father, Jack, being a mining electrical and mechanical engineer in the coal industry, always felt the pressure of the possibility of something going wrong 'on his watch'. Three months before he retired, a cage 'went to the bottom' of the shaft at one of his pits. Fortunately, the fall was not far enough to do other than break a miner's leg.

In November 1987, and coinciding with my 54th birthday on the 10th, we took a short walking holiday in Ullswater in the Lake District. On Monday evening at dinner I raised a glass and proposed a toast to ourselves. Life was indeed good. We had been married for 28 years and were settled in a house in Richmond. Our children were in their early 20s and were successfully getting on with their lives. Jane was Senior Staff Tutor at the Richmond Fellowship. After the successes of launching the Metro and seeing the first stage of Hong Kong's MTR into operation, I was now presiding over a quite remarkable increase in ridership on the Underground. Dangerous hubris, indeed.

## 7.2 The Fire

Eight days later, on the evening of 18 November, we were having dinner with friends in Central London. A phone call came about a major fire at King's Cross station. I immediately went there. In such situations the police and fire services take immediate charge and, for a while, I was a spectator of events. Communications were difficult, unlike today with instant mobile phone access. I met the Operations Director and then the Chairman. He and I quickly agreed that the anticipated pressure from the press would not be handled by our PR man, nor by himself, but by me. Notwithstanding that a heavily increased management load would fall on my shoulders, it was obvious that the press and public would want to hear from the man in charge of LU.

Thus began the most intense period of my life, ending one year later on 10 November 1988, when I resigned just after the publication of the Report on the 'Investigation into the King's Cross Underground Fire' carried out by Desmond Fennel QC, appointed by the government (DoT, 1988).

### 7.2.1 Public Reaction

The national reaction, indeed the international reaction in many cases, was unsurprisingly shocked disbelief. It was the beginning of intense grieving by those injured and relatives of the deceased, including one Officer of the London Fire Brigade. The sense of trauma was, if anything, greater than after Moorgate, and spread further beyond the staff of the Underground — among fire, police and medical services, to say nothing of the trauma among the Underground's own employees.

One week after the fire I spoke to the London Regional Passengers' Committee (Ridley T, 1987) as follows,

> "I must emphasise that, far from cutting back our capital programme there are currently record levels of expenditure on the Underground and we now also have a record number of passengers. This, together with our increased ability to invest speedily and efficiently, means that there is now a strong case for a yet further increase in our capital expenditure budget. However, I must make it clear that we simply cannot put everything right quickly. In spite of all we have achieved in recent years, much of the Underground remains as I described it in 1982 — aged, inefficient and frankly tatty. London's advantage of being the first in the Underground business brings with it disadvantages too."

It also foreshadowed the evidence to the Inquiry,

> "There will continue to be trade-offs in the decisions we take. If we concentrate all our attention on tackling fire for example we are in danger of neglecting problems of safety through lack of capacity — lack of platform space for example."

Lack of capacity had become an obsession with me. Only days before the appearance at the LRPC, I had received a letter from Richard Hope, probably the best railway journalist of my generation, following a long conversation with him. 'Accident statistics and risk analysis are the only firm ground in the swamp when assessing safety', he said, 'and a review of metro fires shows that they are not a major hazard. In 125 years of metro operation, there have been only two incidents that could be described as disaster: Couronne, Paris in 1903, and King's Cross. What generally happens when there is a bad fire below ground is that several dozen people end up in hospital having inhaled smoke. … Two conclusions can be drawn. First, fires are not the most serious hazard on metros, so it would be a mistaken policy to divert resources from preventing collisions, for example, into fire prevention. Second, serious metro fires have only twice resulted in a disaster, in marked contrast to other locations such as hotels and aircraft'. (Hope R to T Ridley, 23 December 1987.)

Awful though the events were, the MD of the Underground could not dwell long on them. There were two and a half million Londoners to get to work and back home the next day, and the next, and the next.

Then there were the press. I had written earlier, somewhat light-heartedly, about public transport bosses being the most hated men in their cities, but now there was a genuine excuse for fierce anger. My observation is that, in the 1980s, the press were still not as vicious as they became by the end of the 'noughties'. Yet I had the good sense to recognise that 'facing the music' was part of the job description. Of course, my experience fell very far short of that of the bereaved but, at times like those, a strong and loving family are an enormous support.

It seems, in retrospect, as though I was on radio and television non-stop for two weeks, trying to do my best. Previous experience promoting the Tyne and Wear Metro stood me in good stead, but there is a big difference between a positive and a negative story. When I look back at videos, I looked drained and tired, though I had not felt that way at the time. Perhaps, a good dose of adrenaline has its uses as well as family support. Much of the press pressure arose because it took a very long time to ascertain how exactly the fire had developed. The press also ran a series of extraneous stories about the Underground that, nonetheless, had to be handled on nearly a daily basis.

Meanwhile, a high priority had to be given to raising the morale of a traumatised work force.

The Report does not describe the life of the staff and managers caught up in a disaster. It is important to learn from such experiences. Of particular importance is the need to keep the organisation working and to support the work force as they face the understandable distress of the public, and cope with their own concerns.

Beyond that a major Public Inquiry had to be prepared for. In addition, I had already set in motion a strategic planning process to allow us to cope with our massively increased ridership. Personally, I was also involved with the DLR as well. This was a major challenge and I was soon convinced that a change at the top management level was necessary, thus leading to the appointment of a new MD reporting to me in the Chairman role. Prior to this I had, in effect, carried three roles, which, on reflection, was most unwise, if not foolhardy.

### 7.2.2 Support during Times of Stress

In addition to family support, I received much encouragement from friends and professional colleagues. Very surprising at first, but not on reflection, was the support received from opposite numbers around the world, most of whom were known to me through the Metropolitan Railways Committee of the UITP. Universally their message was 'there, but for the grace of God, go we'. Subsequently a series of investigations took place in other undertakings, where the possibility of a vast conflagration in an escalator shaft had never been contemplated.

### 7.3 Public Inquiry

The Inquiry was divided into two parts. Part One, opened on 1 February 1988 in Methodist Central Hall, for which there were 114 witnesses, and was devoted principally to eyewitness evidence. In Part Two, that began on 6 April, there were 36 witnesses — including myself and several colleagues; as well as Keith Bright, and Jimmy Knapp of the NUR. The Inquiry moved to Church House on 3 May and public hearings concluded on 24 June, after 91 days. The Inquiry report (DoT, 1988), of nearly 250 pages, is one of the important references in the Bibliography. It included chapters on the ethos, organisation and management of LU, KX station and escalators, an outline of events on the night, the response of LU and the Emergency Services, the management and auditing of safety, station staffing and training, communication systems and fire certification. It also included more than 150 recommendations, very many of which LU itself had proposed to Desmond Fennell, who had been appointed to carry out the Investigation into the fire. Importantly he said, 'there was still no agreement about why the flashover happened so I invited the Scientific Committee to continue work until 31 July. I later extended the deadline to 31 August', as I discuss below.

I make little reference here to the report itself, relying more on the use of Inquiry transcripts. Immediately after the report's publication, one of our barristers wrote a 'brief critique on foreseeability' (Barrister, November 1988) in which he challenged Fennel's logic on that important matter, but which I do not use here. My account, as with other experiences — both good and bad — in other chapters, touches on how it felt to live through the events. None of this, quite rightly, was any concern of Fennel. He wrote what he wrote, and his Inquiry and its report made an important contribution to the government, to London Transport, and to posterity. I regard what I have written here as being consistent with providing 'lessons learned' for the benefit of young engineers.

The report is a long, hard read, but what puzzled scientists, managers, public and lawyers most was — why was there a flashover? I quote Fennel from his Chapter 1.

"The sudden change in conditions between 19.43 and 19.45 when a modest escalator fire was transformed into the flashover which erupted into the tube lines ticket hall, proved immensely difficult for the Scientific Committee to explain. But I am now satisfied that what has been identified and become known as the 'trench effect' is the proper scientific explanation…. The flames begin to extend very rapidly up the escalator trench. In addition, the flames burn more cleanly and smoke emission may fall even though the fire is burning more rapidly. Nearer the top of the escalator, part of the trench flow circulates up over the facia boards, advertisements and ceiling, involving the ceiling paint and producing thick black smoke. In the result the fire was transformed in character by the trench effect causing it to erupt into the tube lines ticket hall at about 19.45 preceded or accompanied by thick black smoke. Without the application of water or fire extinguishers there was nothing to restrain it."

### 7.3.1 A Boost

I was in the witness box from Day 72, after 2 pm on 26 May, until Day 75, after 11 am on 1 June, a total of nearly three days (with a Bank Holiday in

**PHOTO 7.1.  Turn 'em Round.**

*Source*: By kind permission of William Greaves courtesy of *The Times*, London.

the middle). The Underground had a bad press from the time of the fire, and it continued. In parallel, however, there was an amazing outpouring of personal support, much of it by letter to me, but some also in the press. On the day before going into the witness box I had been approached by William Greaves of the *Times*, to come to my office for the purpose of writing a profile. This appeared and was very supportive. Greaves had brought a cartoonist with him, who sketched during our discussion. Thus, under the heading, 'All Signals are for Change — In the King's Cross witness box today, the transport innovator who must put London's ancient Underground back on the rails', (*Times*, 26 May 1988) was a cartoon of me on a station platform, leaning forward with a rope over my shoulder, pulling a tube train into the station, which was named Turn 'em Round.

I had no inkling of what he had intended to write, but it could not have been more encouraging.

"There used to be a custom at board meetings of the LU executive that, whenever a member was caught committing a grammatical offence, he was obliged to pop a coin into the 'swear box' in the middle of the table — the proceeds going to charity. The first meeting presided over by Dr Ridley upon his appointment as MD in 1980 was only a few minutes into its deliberations when a fellow director split an infinitive and was asked to make his contribution. Ridley looked up in some surprise and said, quite mildly, that he would prefer to listen first to the point being made, recalls a colleague. Ridley, who today begins his personal evidence to the official Inquiry into the King's Cross fire disaster, is no domineering disciplinarian. But what the anecdote shows is that the man with the rounded spectacles and slightly disorganised appearance of a prep school master is not easily deflected from his target. He is now recognised as one of the great mass transit innovators of the age."

"To prove that the respect was mutual, Ridley will now reveal his own secret. "I know that they were saying privately, 'Any bloody fool can build a new railway in Hong Kong, but running one that is a hundred years old is a real man's job.' They were right, of course. Not that Hong Kong was that easy, but I knew what they meant." That Far East venture, culminating in the revolutionary MTR, was not his first triumph. With a doctorate in transportation engineering at the University of California and a research stint with the GLC under his belt, Ridley was just 36 when he returned to the county of his birth to become DG of the Tyneside PTE. His brief was to weld together three loss-making bus companies. His achievement was to win £50 million from the government and conceive, design and build the region's 44-station Metro."

"Fellow director, David Howard — now his successor as DG of Tyne and Wear PTE — remembers how Ridley's zeal enabled the authority to leap-frog several cities, including Manchester, whose underground plans were not more advanced. Ridley is hesitant to confirm the story of how he outpaced his trans-Pennine rivals. All I can say is that BR decided to pull out all the stops with a scheme of its own and presented it to John Peyton at the Transport Ministry [in fact it was to Bob Reid, BR's Chairman] one Monday morning — only to discover that Peyton had awarded us our grant on the previous Friday. You have to move fairly quickly in this business."

"Perhaps Ridley should have suspected where his destiny lay when he and his wife, Jane, went house hunting, found the perfect place in Rowlands Gill in County Durham, and discovered that in the garden they had inherited a model railway. There are those who feel that Ridley is still having to move quickly — merely to avoid going backwards, that the story of the last decade has been one of desperate reaction to unprecedented events and that the vast and ancient network — just like any elderly human being — is chronically resistant to change. Confronted with this suggestion, he smiles ruefully. In Hong Kong I was MD of 3,000 people and a brand new company where every decision was a new precedent. In LU there are 20,000 people and loads of precedents. My challenge is to get the ship to change direction — and this particular ship happens to have rather a lot of momentum."

"It is an inbuilt intransigence which Ridley came close to acknowledging when he successfully argued that the new DLR should be independent of the Underground. "What we did not want was a modern railway tied in terms of management, staff and union attitudes to a hundred years of history", he says. Any old organisation is bound to have some old-fashioned attitudes. But John Harvey Jones, the former chairman of ICI uses a phrase in his book, 'Making It Happen', which I believe says it all."

Lesson — Management is about maintaining the highest rate of change that the organisation and the people within it can stand.

"The question on the lips of his staff, however, concerns how much Ridley can stand. While he has been preparing to step into the witness box he has been master-minding the biggest modernisation programme in the Underground's history, possibly spending £3,000 million over the next 15–20 years, presiding over the highest number of passengers — 800 million last year — and acting as chairman of his own brainchild, the DLR. Over an 8 am breakfast in his office on the seventh floor above St James' Park station, he apologises for his desk being covered in mountains of books, files and reports. "I'm afraid they're all King's Cross", he said. He had arrived for work at 6.40 am. Stuart Cole, Senior Lecturer in transport economics at the Polytechnic of North London Business School had seen him leaving for home the previous night at 9 pm."

"It is five years since Ridley forecast the need — now widely accepted — for a massive investment in the Underground. But he steers neatly away from any such political minefield, and prefers to direct his sights on the future. "London is going to be competing with Paris to become the capital of Europe", he says, "and the French won't be hanging around waiting for our challenge." If we are to win we've got to have a city that works. With three million people coming into the central area every day, the existing Underground network can't meet the increased demand. We can't do anything about narrow tunnels and long, meandering corridors, so we must have new lines. That needs not only a management commitment but a political will — and I now believe we have both."

"Ridley's immediate commitment, however, is considerably diverted towards the disaster Inquiry. While the Chairman and MD of LU supports the need for an inquiry, there are many of his staff who feel that the five-month duration is a crippling restraint on the £50 million-a-year business. They point to the comparison with the Moorgate Tube disaster — in which 42 people died — when the inquiry lasted three days. "It's a traumatic time", he says, "but my main job now is to lift the morale of the company." So much lies ahead — and the economic future of London depends on our success."

### 7.3.2 Witness Statement

My Statement to the Inquiry (TMR, 1988a) began with a description of myself – education, brief employment history and professional affiliations. I first presented my approach to the Inquiry,

"In approaching the Inquiry the Board of LUL, and indeed myself as Managing Director, have taken the position from the outset that we would make a full disclosure of all facts that might pertain in any way to the accident at King's Cross on 18 November 1987. We have adopted this policy in the light of Mr Fennell's statement at the opening of the Inquiry that the approach would be inquisitorial and not accusatorial."

I have never regretted that stance, though not everyone was happy with it. The tragedy of the fire could not be changed, but at least it was possible to ensure that appropriate lessons were learned, which is also why this material is reproduced here is such detail.

As for Inquiry being 'inquisitorial', as Fennell said, rather than accusatorial or adversarial, the reaction of myself and most of my colleagues as the proceedings developed was, 'you could have fooled us.' Not only were the proceedings at times distinctly accusatorial, there was another concern. Choices were being made before the scientific investigations into the causes of the fire had been completed. As fully committed to telling the facts as I saw them, it was essential to start to rebuild public confidence in their essential public transport system, without which London would be lost.

The evidence emphasised the age of the Underground and compared its size and ridership with metros in New York, Paris and Moscow. It was important to recognise that London's is the oldest, most extensive, and most complex underground railway in the world. If superimposed on the map of Europe it would stretch across Belgium, from France to the Netherlands. It dated from 1863, and some 80 percent of the system was more than 70 years old.

The challenges faced, and the strategic planning processes we had pursued, are described in Chapter 6. In Operations, changes were being implemented by the introduction of OPO for trains and the introduction of the new UTS. In addition, proposals had been under development to improve the relationship of the organisation with its customers through changed proposals for station staffing as well as customer care and passenger security initiatives. These involved new approaches to recruitment, job descriptions, training and accountability — indeed the whole culture of the organisation.

My Statement went on,

"The Court has indicated that it does not wish to address the overall funding of LRT but will address the allocation of funds and the manner in which decisions were made about the use of financial resources. The Inquiry has already heard comment and discussion about budgets. As a wholly owned subsidiary, LUL is accountable to the Board of LRT. LRT decides on and controls the allocation of all financial resources (from whatever source obtained) to each of its subsidiaries. LRT has control of the investment plans, the annual expenditure budgets for revenue and investment, and fare levels. In turn LRT receives its financial objectives from the Secretary of State for Transport."

"In common with all forms of transport, travel on the Underground is not without risk. This has always been recognised by railwaymen in general, and by our predecessors in LU over the years. ... It has always been thought that the greatest danger of injury or death arose in trains and tunnels rather than on stations. I believe that, even after the tragedy at King's Cross, that remains true and I feel sure that my view would be shared by transport professionals in other parts of the world. However, it has always been recognised that there are safety hazards in stations, whether from overcrowding, bomb scares, floods or fires. What was never foreseen, over all the years, either by the Underground, the London Fire Brigade or indeed by any other body, was the possibility of the disastrous flashover which produced such a devastating effect at King's Cross."

### 7.3.3 Budgets

I was not clear to what extent funding issues should be part of my evidence. 'The Court' did not want to address the overall funding of LRT, no doubt reflecting the government's wishes; and there was no doubt that LRT Board would not be anxious to offer evidence about fierce budgetary discussions that had been taking place with LUL over many months. Such debates are normal and healthy between parent and subsidiary companies in many walks of life. But, were there any 'lessons to be learned' from them?

It was clear to me however that, to the extent that LUL was culpable in not having prevented the conflagration, claiming to have been 'unable to afford to take action for financial reasons', would make no sense whatsoever. But I was also committed to telling the story as truthfully as possible and one issue, which in the event I did not introduce, had troubled me. That was the question of management focus i.e. that senior managers, who by definition must be adept at multi-tasking, were facing strong budgetary pressure and 'making do' had become central to thinking. This seemed a dangerous pre-occupation as the Underground was underfunded and needed much more finance from government or fares in order to maintain its stock adequately.

In the days when GLC appointees were members of the Board, and employment was top of our non-executive colleagues' agenda, not too much of their thinking was addressed to improving efficiency of operation. Equally, after they departed and our masters were now in the Department of Transport, the focus was on the reduction of subsidy. Our internal debates became stronger from the turn of the year 1986/1987. I was not averse to budget savings *per se*. After all, I have recorded the challenge of overcoming inefficiencies within LU. But that, surely, had to take account of the fact that we were now in the middle of a massive increase in ridership.

All project managers are well versed in the original 'fundamental triangle' — time, cost, performance (or quality) — and the need to strike a balance between them, sometimes veering towards one or another according to circumstances. But to focus on one, cost, exclusively seemed distinctly misguided to me and I struggled to convince my colleagues in LRT finance.

The Chairman had agreed that we could base our funding targets, not on cost alone, but on bottom-line figures, i.e. targets which took account of increased revenue generated. I wrote in February 1987, 'Thank you for our helpful discussion on Friday. This memo records my understanding of the outcome. Our approach to emphasising bottom-line funding targets is acceptable, with continuing close examination of costs, as is our intention to examine performance at quarterly intervals. This latter will allow certain existing expenditure to continue until it is found not possible to fund it within the budget, and certain new expenditure to be deferred until it is clear that it can be accommodated within the bottom-line funding budget.' (Ridley T to Chairman, 23 Feb 1987).

However, a letter to LRT from a civil-servant said, inter alia, (DoT to LRT, 4 June 1987)

"The following reservations are entered in accepting the budget and grant claim for 1987–1988 — that the improvements noted in the budget as being required to reduce the deficit in revenue account to £46 million should be secured by reductions in revenue expenditure, and not by reliance on additional revenues beyond those assumed in budget forecasts."

Keith Bright wrote to me, (Chairman to T Ridley, 19 June 1987),

"I confirm that I had agreed with you to go for a bottom-line figure, but the letter from (a civil-servant), which details the Department's qualified agreement to the 1987–1988 budget, gives me little freedom of manoeuvre. However, if LUL can demonstrate significant progress towards achieving cost savings. I will be prepared to go to the DTp to seek their approval to allow some offset of revenue surplus. I cannot do this without evidence that we are doing everything possible to achieve the budget which was approved by the Board for submission to the DTp; and I repeat my offer to sit down and find savings myself."

It has never been clear if the broader message of the need for greater investment in the Underground was getting through, even after the fire. While preparing for my appearance at the Inquiry, I produced a Memo to the Board of LRT, 'LU — Budget Review 1988/1989' that said,

"The past three years have seen a major improvement in the Underground's commercial performance. Unit costs have been reduced by 12.6 percent in real terms, while traffic has increased by 21 percent and revenue has increased £70 million in real terms (+20 percent). As a result, the trading loss has been reduced by £72 million (60 percent) to £48 million. These figures are substantially better than the budgets agreed by the Board. The improved financial performance has enabled the financial targets set by the DTp and, in the main, are better than the annual budgets agreed by the Board?"

"The rise in passenger demand is, however, placing a mounting strain on the system, with the need to increase train services wherever possible and ensure that passengers can be adequately handled within the station environment. In some areas at certain times of day over-crowding is now reaching potentially dangerous proportions. Over the last three years the emphasis has been to reduce costs and I believe little emphasis has been placed on the effect the increased demand has had on maintaining the quality of service, both in terms of train operation and within stations. This is particularly relevant to maintenance of trains to ensure full service every day, station cleaning and effects of graffiti, fraud prevention and passenger security. The Board of LUL believe that the Company is now facing a watershed. The continuation of earlier policies of cost reduction without making additional provision for the need to improve service quality and meet steadily rising demand is likely to result in a serious deterioration in the service provided."

The Memo was withdrawn from the Board agenda. (Ridley T to LRT Board, 26 January 1988).

Although I might have liked to raise financial issues in my evidence to the Inquiry for the benefit of 'learning lessons', it was clear to me then, and still is as I write, that it would have been seen as 'special pleading' and thus counter-productive to other messages I was trying to get across.

### 7.3.4  Change of Evidence

Although many laudatory remarks were made about my performance in the witness box, some of which are included below, I did make one bad error. On Day 73, Friday 27 May, I was examined late in the afternoon by Charles Pugh, acting on behalf of a consortium of solicitors instructed on behalf of the relatives of the deceased and injured. He asked me, "With the benefit of hindsight, do you think that, in the run-up to King's Cross, LUL at the highest level failed to give adequate priority to passenger safety in stations?" I replied, "With the benefit of hindsight, I would say that we did not give it as high priority as we should."

I was appalled when on reading Saturday morning's press. 'Tube chief admits fire deficiencies' in *the Guardian* (28 May 1988); 'Tube chief admits safety shortcomings — there were inadequacies in performance at the highest level' in *the Times* (28 May 1988). *The Independent*, under the heading 'Tube chief admits safety not top priority' (28 May 1988), reported that, in response to the 'with hindsight question', Dr Ridley paused for several seconds before answering, "we did not give as high priority as we should have".

On Day 74 on Tuesday morning, 31 May — it had been a Bank Holiday weekend — I sought leave to make a statement in which I said,

"Perhaps the Court will allow me to refer to the proceedings of Friday when, after the 5 o'clock close of playtime, Mr Pugh asked me a question (at 88B) which I think is important for me to address again. With Fennell's agreement, I went on, Mr Pugh asked me the very important question, "With the benefit of hindsight, do you think that in the run-up to King's Cross, LUL at the highest level failed to give adequate priority to passenger safety in stations?" Mr Pugh had already told the Court that the sun was going down and the temperature was going up and, at that late hour in the day, I fear I misled the Court. I had not realised that at the time, and I did not realise it until I read the press on Saturday morning, which then took me to the transcripts to see that I had been correctly quoted. In *the Independent* in particular the headline was, 'Tube Chief admits Safety not top priority.'"

"I sometimes complain about the accuracy of the press but my reading of the transcripts led me to the view that I had been correctly quoted. May I recall that it was at the end of a gruelling six hour cross-examination. Therefore, I might be forgiven for having misled the Court. What I said in answer to Mr Pugh was as follows (at 88B). "With the benefit of hindsight, I would say that we did not give it as high priority as we should. I went on to answer the following, 'you as a management team were overlooking the immediate concern of passenger safety in stations? You were not giving it sufficient priority as a here and now activity.' I replied, 'I think we recognised that.'"

"I was wrong. I want to assure the Court of this and to assure the public, through the press, that we do and we did give high priority to passenger safety in stations at the highest level — one, from *congestion*; two from *crime*, and three, from *fire*. They are all crucially important. … I have tried to stress that nothing is more important than the problems arising from the very high congestion that currently exists on the system, and I can tell the Court that all the information I have seen suggests that the public is more concerned about crime than about fire, though I am sure they are concerned about fire. I believe the answer I should have given was, 'We do give priority at the highest level to passenger safety in stations, with the benefit of hindsight I believe we have given higher priority to safety problems arising from congestion and crime than to fire, and this was based on our experience of risk.'"

I had long been concerned about passenger congestion and had also developed a considerable knowledge about our passengers' concern about crime, when taking part as a member of the Steering Group for the DoT's report on Crime on the London Underground (Ridley T *et al.*, 1986) and as Chairman of the Working Group. Years later, with further massive increase of ridership, LU are quoted as saying, "soaring population growth could overwhelm the Underground within 15 years, with passengers facing station overcrowding similar to cramming four people in a telephone box. Tube stations are already so overcrowded that some often have to be closed to prevent rush-hour accidents."

I went on,

"Fortunately, I believe I got it more nearly right (at 89E) when I said, and I quote from the transcript, "I am in danger of saying that fire safety is unimportant. It is absolutely not. But we had and have in the centre of our minds this enormous difficulty of congestion. To that extent, but to that extent only, our consideration of fire safety at the highest level was secondary to that of the safety problems of congestion." I believe I was right there and wrong at 88B. I would be grateful if the Court would accept my belief that that is the correct position that I have described and stated, and I trust that it will be given as much publicity through the press as my earlier misleading statement."

There was concern about the over-crowding getting to the potentially dangerous proportions that the 'withdrawn' budget paper to the Board of 26 January 1988 had highlighted. The next day *the Times* (1 June 1988) reported, 'Underground chief corrects evidence on safety priorities', while *the Independent* had, 'London tube chief changes his evidence' It also wrote that, 'part of his role might be taken over by a new MD from outside the organisation.'

In discussion with Pugh about management, I had said, "I have come to the view that we will follow the lead of many companies, both in this country, in continental Europe and in North America, of splitting the role of the Chairman and MD. There will in future be a Chairman and Chief Executive, and he will be supported by an MD who will be driving forward the day-to-day and month-to-month operation of the Underground while the Chairman and CE will be driving forward the longer term, as well as carrying out one of his primary responsibilities, to chair the Board which is a significant task in and of itself." I had not admitted to myself that to do both jobs, when strategic planning was now crucial on top of managing day-to-day operations, plus holding responsibility of the Docklands Light Railway (DLR) development (Chapter 8) was well beyond one man. In retrospect, I suppose the LRT Board should have told me that it was unwise – to say the least. Lesson — Looking back, it was unwise ever to hold both posts as Chairman and MD of LU at the same time.

Pugh asked, "The MD that you are seeking to recruit, may we know whether you are seeking to recruit from within or outside the organisation?" My response was, "I believe that, in any organisation, internal candidates ought to always have the possibility of aspiring to any job, but I would not be at all surprised if the appointee came from outside LUL, indeed outside LRT."

## 7.4 Resignation

At the beginning of November a major reorganisation of the management of LUL was announced with changes of Director responsibilities, and the appointment of ten General Managers for line-based Business Units operating as profit centres within a newly titled Passenger Services Directorate. (LUL Organisation Notice 2817, 1 November 1988) Fortuitously, no changes were necessary when I departed ten days later, though the LUL Chairman/MD relationship did not come properly into effect until Wilfred Newton came from Hong Kong and held the Chairmanships of both LRT and LUL.

Prior to the publication of the Inquiry report (DoT, 1988) we were fore-warned when it would become available and I was anxious to provide copies to the LUL Board so that we could immediately prepare comments and be ready to make comments to the press, who would undoubtedly approach us. But life did not go according to plan or, more accurately, that was not the plan that the government had in mind.

### 7.4.1  A Memorable Interview

On Monday, 8th November, I was invited to go to see the DoT Permanent Secretary in his office at 2 pm. I expected to be given a copy of the report so that I could return to 55 Broadway to discuss its findings with my senior lieutenants. How utterly naïve. He gave me a copy and told me that I had two hours (!) in which to read it. I would be given a room in which to do so. After that I would be ushered into Channon's office, the Secretary of State for Transport, to discuss it with him. Keith was already doing exactly the same. We would not to be allowed (*sic*) to meet each other for any discussion.

I should have said, 'This is quite ridiculous', but was taken aback and realised that was exactly what was intended. I went to see Channon. It was awful. He asked me what I thought of the report, and then mumbled that it was 'all very difficult'. He did not have the cojones to say, "I want your resignation".

I said I would give him a considered view when I had had time to read the document properly. When I met Bright I learned that he had resigned and issued a statement.

"As was disclosed at the Formal Investigation, after the King's Cross fire I offered my resignation as Chairman immediately as a matter of principle, being the person with overall responsibility for the organisation and all of its subsidiary companies. The Secretary of State for Transport asked me to carry on because of the importance of the immediate tasks ahead. As a result, during the past year I have devoted most of my energy and attention to seeking to ensure that the lessons of King's Cross have been well and truly learned and that actions necessary to avoid a recurrence of such a tragedy should be vigorously pressed ahead. Having accomplished the task set me by the Secretary of State, with the impending publication of the Fennell report, I have told Mr Channon that I feel now is the appropriate time to hand over to a new Chairman." (Chairman to staff, 7 November 1988)

I wrote to the Secretary of State, 'You have given me time to have an initial look, confidentially, at a copy of the Fennell report on the King's Cross fire, in advance of publication. It is an emotive document detailing the events and background to that tragic incident. It makes a number of criticisms of London Underground both before and during my tenure of office. At the same time it acknowledges LU's outstanding safety record and the fact that the King's Cross disaster was unprecedented in the experience of LU or indeed of the emergency services and the Railway Inspectorate. You have asked me urgently to consider my reaction. Obviously, in order to give the report the consideration it deserves, I and my colleagues need time to study it in detail but I have little doubt that the recommendations are sound and indeed a large number were proposed by LU and are already in the process of being implemented' (Ridley T to P Channon, 9 November 1988).

It seemed unprofessional, to say the least, to press me for an urgent reaction to the report when it was clear that he and his civil servants had had a copy for very many days, and only let me see it, very briefly, just before publication. Clearly the game-plan was to get me out of the way before he stood up to make his statement in the House of Commons. I reserved my position.

Meanwhile, the Boards of both LRT and LUL had written to Channon, seeking my continuation. LRT (by now under the chairmanship of Neil Shields) said that the Board believed that the Secretary of State should not ask for or accept my resignation. They said that the safety record of LU was outstanding, a situation which did not occur without safety being addressed and treated seriously. My expertise, particularly as a railway builder, was held in the highest regard by the Board and the development of the railway and the railway organisation was at such a point that to lose my abilities and contribution, particularly at the time, would be counterproductive and totally against the interests of LRT and LUL.

LUL Board said that, having regard to my outstanding contribution to LUL as a leader, manager and strategic thinker, the Board of LUL confirmed its complete confidence in me as Chairman. It believed that the loss of my leadership and railway experience would seriously weaken LU's ability to deal with the problems which it faced in meeting the needs of London both then and in the future'. However, by the evening of 9 November (12 months to the day after our Ullswater dinner), I decided that I would resign. It became apparent that, notwithstanding the many hurrahs for me, the circumstances required someone to be seen, publicly, to 'pay the price'. Moreover, the Minister simply dared not have me still in post when he stood up in the Commons. After much reflection I decided that it was, in fact, for the best. It was at least arguable that, if I had soldiered on, I would no longer have had the influence necessary to fight for all the public and political support the Underground needed

By 5 am the next morning, my 55th birthday, there was already a crowd of reporters and cameramen standing waiting across Church Road, Richmond, opposite our house, number 77. I debated how I could avoid them. Wisely I rejected trying to hop over the garden wall at the back of the house — that would have made a very poor picture on the front of the Standard. The Standard (9 November 1988) did have me on the front page however, 'Ridley's last ride in the driver's cab.' It said,

> "As long as you are chairman of London Underground you have always got an escape route if people bother you on the way into work. If people keep asking you if and when you are going to resign you don't have to put up with them to the end of your journey like any other commuter. You simply keep walking right to the end of the train and climb in with the driver. It may be dark for reading the *Times*, but it gives a little time for reflection on how you will answer the questions at the other end. For the record Dr Tony Ridley was saying no comment when he appeared at his Richmond front door this morning. Briefcase in hand, he smiled and said nothing about reports that he, like Sir Keith Bright, had tendered his resignation. It's a seven minute walk down the hill to Richmond station. Today Dr Ridley made it in five. Then it was a brisk stride along the train and into the driver's cab, where he stayed until St James's Park."
>
> "On the way he let it be known it was his birthday and why didn't I ask him what he had been given'. So I did. "Lots of lovely presents from my family", he beamed and strolled on, choosing not to say whether he had read the Fennell report or what his future employment plans might be. Whether they held a District Line train for him or whether his timing was just impeccable we will never know [it was neither — I was fortunate in the timing]. However, he never broke his steady stride through Richmond station booking hall and on to platform five. There were scores of seats available on the train waiting there, but he kept on past them to the head of the train. A discreet flash of his official pass to a slightly puzzled looking driver, a signal turned green and the doors slid shut."

LUL then issued a press release.

"Dr Tony Ridley, London Underground's Chairman and Chief Executive, has resigned this morning. He said, 'I have had the opportunity to take an initial look at a copy of the Fennel Report on the King's Cross fire to be published today, and to discuss it with Transport Secretary Paul Channon. It is an emotive document detailing the events and background to that tragic and unprecedented incident. It makes a number of criticisms of LU both before and during my tenure of office, many of which I recognised at the Inquiry. In leaving, I believe it is vitally important at this time that the travelling public recognises the fact that the Underground is one of the safest forms of transport, with an exceptional record prior to King's Cross. I am resigning with much personal and professional sadness because of my natural desire to lead the future development of the Underground. I am, proud to have led LU from a low ebb in the early 1980s to its present position of record traffic, a sound financial base and a clear strategy for the future. I also take pride in the action programme which the company has instituted since the fire and which forms a large part of Mr. Fennell's recommendations.

In accepting my resignation the Secretary of State was kind enough to pay tribute to the major contribution I have made to LU and the DLR in the past and shaping the future of London's transport system. I should like to take this opportunity to express my deepest sympathy to those who were injured or bereaved by this appalling and unprecedented tragedy, and my thanks to all my colleagues and staff in LU who have worked so hard to prevent possible recurrence.'" (LUL Press Release, 10 November 1933)

### 7.4.2 The Press

The press did not, to say the least, make happy reading, being properly critical. However the *Times* also said,

"Their (Bright and Ridley) resignations were appropriate as a matter of honour — something rare in public life today. ... But their departure holds risks. In the first place, LT has been deprived of two particularly well-qualified executives. Sir Keith has pioneered policies of high efficiency, service to customers and involvement of the private sector on LT should soon bear fruit. Dr Ridley is an underground railway engineer of world renown. In the second place, their resignations might prompt the erroneous conclusion that they alone can be held responsible for the disaster. This could have the lamentable effect that the many other points made in the report of the public inquiry will not receive the attention they deserve" (11 November 1988).

It was particularly interesting to read the Guardian. The principle message was that the report had 'thumped safety' on to every company board agenda. Although the editorial did not say so, it was reinforced by the context of other well-publicised tragic accidents at about the same time — the Herald of Free Enterprise ferry at Zeebrugge in March 1987 and the Piper Alpha oil platform in the North Sea in July 1988, were themselves followed later by the Marchioness pleasure-boat disaster on the Thames in August 1989.
*The Guardian* also said,

"London Underground is the safest way to travel in the city. It saves lives. It is also, increasingly, the best environmental way, saving lead-free lives too. Its passenger usage has gone up. It is dirty and overcrowded, and struggling — successfully — to generate the money it needs to build, refurbish and expand. Political doctrine has little to do with that struggle. It is successful in attracting passengers that is the market force upon politicians of any hue. The departed Dr Ridley made his reputation building undergrounds in Tyneside and Hong Kong. They, among other things, saved lives. The far-reaching plans for London, announced only a few weeks ago, will do that too, as well as make the capital a more civilised place to live." (*Guardian*, 11 November 1988)

Richard Hope said in the *Sunday Times* (13 November 1988),

"Joining the Underground was a job for life. Pay might be poor, but promotion was assured through seniority, not performance. Young people with talent became frustrated and left. Fresh talent was rarely sought; it was tacitly assumed that no outsider could possibly grasp the intricacies. This was the Underground that Ridley found when he took control in 1980, straight from building Hong Kong's highly successful underground system. The culture shock was formidable."
"Productivity on the Underground had fallen dramatically in the 1970s. Restrictive practices were endemic. For decades investment had failed to keep abreast of decaying assets. When Ridley tried to set matters

right, his bosses in the Labour-controlled GLC blocked every turn. True, they had progressive views on such innovations as fares which set the stage for the 60 percent surge in passengers since 1982. But job protection policies and the left's fundamental assumption that all men and women are equal when it comes to promotion reinforced the deeply ingrained public service ethos of the Underground."

"Bright burst upon this cosy scene like a demon king in a pantomime, following the government's takeover of LRT in 1984. ... Above all else he believed in — and applied rigorously — the stultifying discipline of budgets. Nicholas Ridley, the then transport secretary, was overjoyed. He told LRT to halve its subsidy in three years. Bright did it too, promising more."

"As the need for subsidy plummeted, more cash was released for investment, first to smarten up stations, then to cope with the avalanche of passengers who switched from worsening traffic jams on the streets above. ... Well aware that the Underground was on a knife-edge — though his nightmare was overcrowding, not fire — Ridley of the Underground pleaded with Bright to ease the pressure. He got his answer in October 1987. In the first six months of the 1987–1988 financial year, costs were £4 million over budget. Naturally, carrying more passengers had used more electricity, tickets and the rest. But revenue was up £8m. Bright would have none of it. [That is not correct — TMR] Budgets were sacred; traffic levels irrelevant. Another round of cuts was ordered."

"For better or worse, both men are gone. Ridley will not grace the dole queue for long. At the budget battle's height, Alistair Morton, Eurotunnel's Co-Chairman, urged him to take charge of the project. At the time Ridley said he was turning it down "because I haven't finished the job I came to do in London." That all went up in smoke the next month. Perhaps his greatest failure was not standing up to Bright, and threatening to join Eurotunnel."

I had left LT forthwith, on 10 November, but retained my office until I had time to clear my very many books and papers, some of which have given me the opportunity to refresh my memory in writing this chapter.

With Jane's guidance I had eight sessions of professional Counselling with one of her colleagues to help me cope with the pressures of the work and the deep sadness and disappointment that my time at LU would end with such a tragedy. I was also much supported by professional colleagues, and friends. Lesson — Seek professional Counselling if you feel it is necessary and accept support when given.

### 7.4.3 Welcome Support

After leaving it was surprising how quickly I was approached again to join Eurotunnel in an executive position. But I did have time to put together a document that bolstered my spirits. I have already said that I received an outpouring of personal support — from friends, relatives, national and international colleagues, current and previous board members, my own staff, trade unionists and many others. It was extraordinarily complimentary, embarrassingly so. I came to believe they could not all be wrong in their assessments. I assembled 17 pages of it (Ridley T, 1988b), some of which, immodestly, I reproduce here. In a letter to *the Independent*, Terence Bendixson (18 November 1988) wrote,

> "Mr Ridley, who created Tyneside's metro and built Hong Kong's, is a man with precisely that vision that Mr. Glancey (in a previous letter to the Editor) calls for. Mr. Ridley is of the stature of Frank Pick. Unlike many people, he knows that cars, far from killing undergrounds, double their potential. This knowledge enabled him to persuade the DoT, BR and the Underground to work together on railway investment for London. Meanwhile he was transforming the Underground management and also, thanks to his skill in obtaining government financial support, getting approval on a huge scale for badly needed trains, signals and stations. Finding a replacement for this exceptional man is by far the greatest challenge facing the board of LRT. But, make no mistake, Tony Ridley has shown the way."

A former LT Board member wrote, 'I am clear that you have been the most effective Chief Executive of the London Underground since Frank Pick.' That was from Anthony Bull, one of my mentors, the man who had persuaded me that it was alright to leave Tyne and Wear to go to Hong Kong. I blush, but I could not have been more heartened. And from an LRT Board Member, 'The irony is that we have lost the services of the person most able to do something practical about London's transport crisis, just when it is most needed and just when you had got yourself into a position where you could concentrate on what you do best.'

Sir John Drinkwater QC, the barrister who represented the London Fire and Civil Defence Authority wrote, 'a very brief note to congratulate you on the way in which you approached and carried out your huge task on giving evidence to the Inquiry. To me, and I would judge to all your friends, you made very plain your intellectual stature and your unshakeable integrity. One hears the Inquiry gossip and there is no doubt that you impressed the Inspector, the Assessors and the Bar as having given a brilliant example of how to give evidence' (Drinkwater J to T Ridley, 3 June 1988).

Some comfort came from many of Henderson's statements to the Inquiry, that the corporate view of LU was most clearly and eloquently expressed by myself; that I was a man who was so obviously a leader that that was an impression which must have been gained by most people who were attending in Court, a leader of men; that I expressed my views with an intellectual clarity which was a pleasure to hear.

> "Travelling by Underground train and using Underground stations has been an exceedingly safe means of public transport. Very few people indeed have lost their lives this century by such travel, and only a handful have done so as a result of fire until the King's Cross disaster. None before the King's Cross disaster had lost their lives from an escalator fire, although there had been a number of such fires. For this LTE, LRT and LUL deserve full and proper credit, and these opening submissions must be seen against that essential background."
>
> "We derive great assurance from the tone and substance of the evidence of Dr Ridley as regards his determination to ensure that the lessons that are fit to be learned are truly learned. This is no more than an impression and it has to be seen in the context of undoubted recognition by LUL, to a far greater extent than was ever recognised by LTE or its masters, of the need for radical rethinking of some of the most basic tenets of the underground railway management. Safety was not ignored. The safe carriage of thousands of millions of passengers did not happen by accident but design. For this LU and its predecessors deserve full credit. We are unhappy that our commendations in this respect have not been as widely reported as our criticisms. Such is the way of the world. We hope that on this occasion a fair balance will be maintained by including in our closing submissions the words which we used when recognising the phenomenally successful record of LUL and its predecessors in terms of avoidance of major accidents and multiple fatalities, especially multiple fatalities from fire, over many decades, with very few exceptions unrelated in almost all cases to fire. All our criticisms need to be seen in that context. It is very important that people should not lose confidence in their use of the system if, as is the case, as a whole it is safely managed and safely run" (Henderson, Day 41).

Even Fennell, whose report poured immense criticism on LUL said, 'It is clear from what I heard that London Underground was struggling to shake off the rather blinkered approach which had characterised its earlier history and was in the middle of what Dr Ridley described as a change of culture and style'. 'These proposals for change, many of which were in progress before the disaster, are far-reaching and I do not doubt the commitment of Dr Ridley in seeing them through.'

Professor Bernard Crossland, one of the Inquiry Assessors and one of the great engineers of his generation, who had watched me throughout my involvement in the proceedings, said, 'I always feared that some heads would fall, but for myself I was particularly impressed with his response to the Inquiry. Indeed, I said to the Inspector that if I was a young man again I would have liked to work with Dr Ridley.' Bernard's obituary in *the Telegraph* (23 January 2011), said of his role in King's Cross, 'As head of the scientific committee investigating the cause of the fire, Crossland led a painstaking operation which involved both forensic and experimental scientists and made a significant breakthrough, discovering an effect hitherto unknown to science.'

Later, Bernard caused some consternation when he was quoted calling Public Inquiries into such disasters time-consuming, expensive and inefficient. By the time the report of the Inquiry was made available the general public and the politicians have lost interest. It is not for me to agree or disagree with him. However, ideally Public Inquiries about accidents should not begin until any scientific investigation, for causes, has been completed and reported. As Fennel wrote in Chapter 2 of his report, 'the public hearings concluded on 24 June 1988 after 91 days. There was still no agreement about why the flashover happened and so I invited the Scientific Committee to continue work until 31 July 1988. I later extended the deadline to 31 August 1988 to enable further experimental work to be undertaken and to allow the parties sufficient time to make their final submission on technical matters' — a mere 8 weeks before the publication of the 247 page report! Lesson — Ideally Public Inquiries about accidents should not begin until any scientific investigation, for causes, has been completed and reported.

I took comfort from correspondence from members of my staff though, no doubt, they were sympathetic to me for standing up and taking all the flak.

From an LRT manager, 'I am writing to say how saddened I was by your resignation — and to express my admiration for the extraordinary fortitude and determined good humour with which you have come through this gruelling year. It must have taken a superhuman effort to achieve that, while dealing with the enormous additional work required to satisfy the Inquiry, and the continuing management of the Underground. When you joined, one had the impression within weeks that it was acquiring a new energy and sense of purpose, and it is difficult to realise what a transformation has been achieved, so much that the improvements in efficiency and service have become taken for granted. And you achieved that, not just through energy and vision, but through a warm and sympathetic approach to staff at every level. We are all indebted to you, not just for giving us an organisation which we can begin to be proud of, but for the personal example of the way business should be conducted and values upheld.'

To each of the personal letters I sent the same typed reply, with longhand additions to each. I wrote,

> "So many people have telephoned, spoken and written to me (and Jane) it is difficult to know how to say thank you. I can only say that, while I perhaps expected friendship at this time, I have been overwhelmed by the outpouring of regret, support and encouragement as well as so many very, very flattering remarks about integrity and accomplishment. I am glad that our friends have recognised our own trauma and the tremendous load that the family have helped me to bear, not just recently, but for more than a year. I frankly expected withdrawal symptoms on Friday, 11 November. In fact I experienced a tremendous sense of relief with the lifting of a burden. I am sad that I shall not be following up so much of what I have set in motion. However, it is comforting that my colleagues recognise that London Underground has survived when it might have disintegrated. There is a new strategy in place with a massive increase in funds committed, a new organisation in place, and a new rail strategy in London well on the way. So we are in good heart, particularly with so much love and support around us" (Ridley T, December 1988).

I also had some supportive words from members of the press, but not all of them. My favourite satirical magazine, *Private Eye*, had a go. It wrote (25 November 1988),

"If maintaining good relations with the media had been a criterion for holding on to his job as LU's chairman, Dr Tony Ridley would not have lasted as long as he did after the King's Cross fire. The day after the tragedy he accepted an invitation from *Thames TV* to be interviewed on its regional news programme. However, his behaviour at the studios did not endear him to the hacks. First he demanded that he be interviewed at the start of the programme. This, it was felt, could not be granted as the viewers would be expecting to see reports about the fire and its aftermath first. Then the doctor complained furiously that he had been shot in profile, which was not his best side. Bemused staff found it difficult to grasp that this was the principle concern of the Underground's chairman the day after a fire which had claimed 31 lives. Dr Ridley then left the studio declaring that he would never again appear on a *Thames* programme. Hopefully, in view of his recent resignation, he may not be asked."

One of those journalists I had met during my strategic plan media programme was Rob Kirk, the Editor of *Thames TV News*. Two weeks later he replied (Kirk R to *Private Eye*).

"God knows, Dr Tony Ridley (ex-boss of LU) has got enough problems without your totally inaccurate report about him and *Thames News*. For the record, Dr Ridley WAS interviewed by *Thames News* the day after the King's Cross disaster, but NOT — as you suggest — at our studios. Indeed he gave us THREE interviews — one live and two on tape. At NO time did he insist on being interviewed 'at the start of the programme'. Nor did he 'complain furiously' about the way he was shot; as far as I'm aware he didn't complain at all. As for his alleged 'vow' never to appear again on a *Thames* programme — well, he did appear many times after the tragedy, including an extremely frank interview with *Thames News*, broadcast on the day of the Fennell Report. I suspect that the 'vow', like everything else in your report, is a fiction."

I dropped him a line to thank him, saying that it was good to know that the record was put right just occasionally. (Ridley T to R Kirk, 21 December 1988).

These compliments are not included as self-congratulation, although they were comforting. But when these experiences are published I am anxious to leave some messages for young members of the profession. Very few of them, happily, will face major disasters during their career but, if they do, perhaps they will provide them with some guidance. However, there is a larger point — if, in the course of your career, one of your colleagues and or friends runs into deep water, a communication of support is enormously welcome.

## 7.5 Thereafter

Naturally, the press will never give up, nor should they, although since 2012 there is much discussion about what limits should be put on the press, and what are their ethical standards, if any. But shock/horror will always be with us and my successors have had to live with it. Three years after my departure, the *Evening Standard* was writing, 'Error after error as the Tube tried to cope with its biggest emergency' (13 August 1991), followed the next day by 'Lapse in safety rule led to new Tube blaze' (14 August 1991). It editorialised, 'With a perverse irony, hundreds of passengers were stranded in tunnels on the Underground yesterday after a fire in a storage space, on the very day when details of a report on the stranding of 7,000 people below ground last February became public. ... The culture of sleepy mismanagement is so deeply rooted in London's official transport infrastructure that there seems to be little hope of any fundamental change for the better.'

*The Standard* (20 June 2000) kept on and on, saying 'Mr Tunnicliffe declares he is leaving the Underground in an 'overwhelmingly better' state than he found it. This will be news to passengers who have seen a deterioration in service — delays and breakdowns far more damaging than they were in the late 1980s — coupled with a doubling of average fares.' Sad, and very unfair, but it was not until years later, after the transport triumph of the 2012 Olympic Games, that the press started to recognise the fact that London has a rather good public transport system.

## Bibliography

Barrister Critique on Foreseeability (November 1988).
Chairman to T Ridley (19 June 1987).
Chairman to LRT staff (7 November 1988).
Department of Transport (1988), Investigation into the King's Cross Underground fire, HMSO.

DoT to LRT (4 June 1987).

Henderson R, Extracts from Inquiry transcripts, (Day 41).

Holloway S (1988), *Moorgate – Anatomy of a Railway Disaster*, David & Charles, Newton Abbott, UK.

Kirk R to Private Eye (23 December 1988).

LUL Organisation Notice 2817 (1 November 1988).

LUL press release (10 November 1988).

Private Eye (25 November 1988).

Ridley T to Chairman (23 February 1987).

Ridley T (1987), The King's Cross fire, Statement to London Regional Passengers' Committee, London, 25 November.

Ridley T to LRT Board (26 January 1988).

Ridley T to P Channon, Minister for Transport (9 November 1988).

Ridley T (1988a), Witness Statement, King's Cross Inquiry.

Ridley T (1988b), Extracts from articles and correspondence (unpublished).

Ridley T *et al.* (1986), Crime on the London Underground, In *Report of a Study by the Department of Transport in conjunction with London Underground, the Home Office, the Metropolitan Police and the British Transport Police*, HMSO, London.

## Press

Bendixson T, *Independent* (18 November 1988).

*Evening Standard* (9 November 1988).

*Evening Standard* (13 August 1991).

*Evening Standard* (20 June 2000).

*Evening Standard* (14 August 1991).

*Guardian* (28 May 1988).

*Guardian* (11 November 1988).

*Independent* (28 May 1988).

*Independent* (1 June 1988).

Hope R, *Sunday Times* (13 November 1988).

*Times* (26 May 1988), All the signals are for change, William Greaves article.

*Times* (28 May 1988).

*Times* (1 June 1988).

*Times* (11 November1988).

*Telegraph* (23 January 2011).

## Personal correspondence

Drinkwater J to T Ridley (3 June 1988).
Hope R to T Ridley (23 December 1987).
Ridley T to R Kirk (21 December 1988).
Ridley T response to personal letters (December 1988).

# 8

# Docklands Light Railway

**Did the 'Toy Train' contribute to winning the Olympic Games for London? Can a creative politician achieve more by encouraging engineers?**

Some lessons

- To enable speedy development it may be wise to develop separate management structures.
- Adding capacity during construction is complex.
- To 'make things happen', it is sometimes necessary to grasp opportunities as they occur.
- Creating new transport links creates opportunities for growth.
- My Olympic role did not need so much planning, engineering, or even political skills, but needed interpersonal skills.

## 8.1 DLR from 'White Elephant' to 'We Need More Capacity'

The Docklands Light Railway (DLR) story began in 1980 during my time in Hong Kong. By coincidence Nigel Broakes, who would become the first Chairman of the London Docklands Development Corporation (LDDC), was in Hong Kong as the Chairman of Trafalgar House, owners of Cunard. He held a cocktail party on board the QEII in Hong Kong harbour to which I was invited as the MD of the MTR.

In conversation with him about the Docklands development I said that I supposed he would have to have a transport spine down the middle of the development — a little railway. "Oh no", he said "there's no way the government would pay for something like that, there's no chance of that happening". I thought that was very strange and, if he was going to make the development happen, he would need a rail line. We only had a brief conversation, but it drew my attention to what was the current thinking about the development of Docklands. At that stage there was no prospect of me becoming involved in any way.

I joined London Transport in September 1980 as MD of London Underground. In 1981, the LDDC was established. Within the LU the Development Director, Bob Dorey, together with David Dell, a senior civil servant from the Department of Trade and Industry, and George Curry who was the Executive Director of the Railway Industry Association, had been having meetings about how Light Rail might be developed in London. This was very unfashionable as virtually no-one associated with London Underground at this time thought that it was of any consequence.

In particular Michael Robbins, my predecessor but one as MD, would have nothing to do with the technology. Also, within UITP, there was competition between metropolitan railways and light railways and Robins had been the Chairman of its Metropolitan Railways Commission. I, of course, had become President of LRTA while I was still at Tyne and Wear.

During 1981 Paul Cautley, Chairman of the D-Group, introduced me to Reg Ward, the first Chief Executive of LDDC. We had a breakfast meeting at which we discussed my interest and his wish to develop a transport system with new modern technology within Docklands.

### 8.1.1 No More Underground Lines

Meanwhile, there had long been a desire to extend what was then known as the Fleet Line, and later the Jubilee Line, through the east end of London and crossing the Thames five times. This would have gone through the Isle of Dogs. In 1980, Parliamentary powers were obtained by LT to construct the line from Charing Cross to Woolwich and a considerable amount of money was spent in safeguarding the line through the City. However, the powers were allowed to lapse.

This attitude was understandable to some extent. Ridership on the Underground had been falling for over 30 years, since shortly after the end of World War II. That was the principal challenge to me as a new MD. Ridership had fallen from 720 million journeys in 1948 to 498 million in 1982. What should we do about it?

The plan for Docklands at that time was for low-density development, which could not have justified the expense of building an Underground line. Nonetheless, Reg Ward and I had no difficulty in believing that Docklands might justify a small, relatively low cost, transit system as a backbone. It should be pointed out that this was well before the enormous Canary Wharf towers, that we all know today, had been proposed. Using some of the lessons from Tyne and Wear we determined that the only affordable way was to use existing disused railway right-of-way for parts of the route. The earlier intention for Docklands was to fill in the old docks to create more land. Once it was decided to retain the docks the elevation of the railway over part of its length became, happily, inevitable. The picture of the brightly coloured DLR, the 'toy train' as it was disparagingly called initially, has long since become an icon for the whole of Docklands.

During the whirlwind of the early planning a number of options were considered. Reg, however, had an obsession for an automated railway — he wanted a 21$^{st}$ century system and a 21$^{st}$ century image. He had met Canadians, from the Lavalin company who had built the automated 'Skytrain' system in Vancouver, led by Kirk Foley (known by us all as Captain Kirk). Ward wanted their 'whizz-bang' technology.

Thus LDDC, in the person of Ward, wanted a modern railway, and I was enthusiastic to make it happen. The GLC was also involved, partly because of land ownership but also because it fell within the area of its overall transport responsibility. As we have seen, shortly after the establishment of LDDC, Labour won a GLC election in 1992 and Ken Livingstone's '*coup d'etat*' took place.

Dave Wetzel became Chair of the Transport Committee and his big contribution, in this matter, was to ensure that his political caucus agreed to GLC working with LDDC, particularly on the question of transport. The left were opposed to the LDDC, which they saw as a Tory-imposed development corporation riding rough-shod over the local councils and, indeed, local people. At that stage we, London Transport, were not in charge of the project. We were effectively the agents of two joint clients — LDDC and GLC. Later the project became a creature of a tri-partite client, with LT joined in. Earlier, decision-making between the two — LDDC and GLC — was extremely complicated with Ward acting for LDDC and Audrey Lees, a GLC officer, acting for them.

### 8.1.2 Go-ahead for a Low-Cost Railway

A scheme was worked up by a very small joint LT/GLC team, involving consultants Kennedy and Donkin, for an inexpensive system with a preliminary cost estimate of £65 million (excluding the cost of inflation). A request was made to government for financial support. To our surprise we learned that Michael Heseltine the Secretary of State for the Environment, (who I regard as the 'father' of Docklands and the best Tory prime minister we never had), had persuaded the then Financial Secretary, Leon Brittan, to agree to finance the scheme. It was well known at the time that the Department of Transport under David Howell was initially not in favour of the project — the cost-benefit analysis showed that only half the project could be justified by the transport benefits. However, it was estimated that more than 9,000 extra jobs would be created by the light rail scheme. This led to the agreement that the cost would be shared equally between Transport and Environment. The shortcoming was that some civil servant, without consulting us, had taken our cost estimate of £65 million, added a percentage for inflation over the construction period, and arrived at an all-up figure of £77 million. Take it or leave it, said Heseltine and, by the way, when you let the contract it has to be on a package basis. In other words one company or consortium, after competition, was to design and build the whole system.

In October the Under Secretary at DoT, wrote to LT's Chairman, Keith Bright,

> "I know you will be delighted to hear that the Secretaries for Environment and Transport are announcing today that the government has approved the Light Rail Project for London's Docklands, linking the Isle of Dogs with the City and Mile End'. The government had agreed to make the necessary funds available to enable LDDC to meet 50 percent of the cost of the scheme. The DoT designated the project as one of regional and national importance and accepted GLC's share of the project cost for grant without prejudice to other GLC capital projects." (Bridgeman to LT Chairman, 7 October 1982)

A copy of the Departments' press release (DoT, 7 October 1982) was enclosed.

> "The Government has approved a light railway for London's Docklands, linking the Isle of Dogs with the City and Mile End. The scheme will cost

£65 million and, subject to Parliamentary approval work could start early in 1984 and be completed by 1987."

Wow — this made the T&W Metro decision-making look like slow going!

The government had been persuaded that the project would be a major stimulus to the regeneration of London's Docklands by making the area a better place in which to live and work, and make Docklands a much more attractive prospect for investment by employers and developers. They also saw that a light railway would be quicker to build than a conventional link, cost less to construct and would be cheaper to run.

The terminus of the new railway would be near Tower Hill underground station. For part of its route to the southern end of the Isle of Dogs it would use a viaduct formerly used by British Rail. The Mile End branch would run from Poplar to Mile End underground station, following a disused BR right-of-way to Bow Road, thence along the street to Mile End. London Transport would table a private Bill in Parliament in the 1982/1983 session to obtain the necessary powers for the City — Isle of Dogs section.

In November, June Bridgeman wrote to Jim Swaffield, Director General of GLC,

> "to record formally the Government's commitment to allocate resources to the DLR. .... I should emphasise that the Government's commitment is to a DLR as described in the June 1982 report and built within the cost and time parameters set out in that report. It is therefore essential that expenditure on the project does not exceed the overall total of £77 million and that it is completed within the agreed timetable; the project should take full advantage of existing technology but should not incorporate technical solutions which would significantly increase the risk of cost overrun or compromise the railway's completion date or reliable and economical operation."
> (Bridgeman to Swaffield 24 November 1982)

During 1983 a street running option to Mile End was reviewed. LDDC were strongly against the prospect of a system with overhead catenaries, essential for a street running railway, running through the development area. An alternative, much better, route to Stratford again using former BR tracks, was developed and opened up the prospect of an automated railway with ground level current collection.

There was no great enthusiasm among the public for the scheme. Quite apart from the objectors to the Docklands development overall, the attitude tended to be — why build this railway through a derelict part of London. In the minds of many it was a complete 'white elephant' — interesting in light of later criticism that its capacity was totally inadequate.

It was now necessary to obtain Parliamentary powers to construct the system and then let the contracts. There were three consortium leaders who bid — Metro-Cammell (led by Tony Sansome), GEC (General Electric Company whose MD was Arnold Weinstock) and the Canadian Lavalin (led by Kirk Foley). Ward was hopeful that the Canadians would win but, because it had been agreed that LT would build and operate the system, it was an LT decision. Ward was a member of the procurement committee. His strength was that he was unconventional — necessary to create a dramatic new development, but somewhat difficult in that he would be with us during negotiations with all three groups, and then spend the evenings with the Canadians. Paul Cautley represented Lavalin and, when we travelled to Vancouver to consider their already-operating system, it was Paul who made the arrangements.

In the event the successful consortium was GEC-Mowlem, with GEC providing the railway and all the sub-systems, and Mowlem the civil engineering contractor. Roger Sainsbury, a future President of ICE, was in charge of the project for Mowlem.

## 8.2 Implementation

A major decision was now necessary about how to manage the DLR project. The MD of LU was now, on behalf of the Board of LT, responsible for the development of a brand new automated railway. This meant that effectively I had two major complex jobs. If the Underground organisation was used to develop the DLR, first it could take forever and also it might not turn out to be a light railway. Indeed the Tyne and Wear Metro veered, over its development phase, towards many of the characteristics of a heavy railway.

The Underground was both experienced in, and committed to, underground railways but not to modern light railways. A separate organisation became the best option, very modest in size, with staff seconded from the Underground and also supported by project management consultants, led by

Mike Nichols who had been involved as a consultant with both the MTR and LU. Not everything went well with the project but I am very clear that the decision to create a separate management team was correct.

Lesson — To enable speedy development it may be wise to develop separate management structures.

In March 1983 a paper was presented (Ridley T, 1983) to the LT Executive Committee, 'DLR Project', in which the background to the project was described and, *inter alia*, the relationship between LT and the two clients, control by LT, and LT's internal arrangements, and said,

> "It has always been the intention that the DLR project should be separate from the railway business and seen to be so — so as to avoid diverting departmental resources from their main tasks of running the Underground; so that it can develop its own management, operating, staffing and negotiating procedures appropriate to a low cost, low density system; so that it can be separately accountable; and so that it can buy any supporting services it requires from LT or elsewhere."

There was already a Project Manager reporting to me (Bill Clarke until he became Operations Director of LU, when Cliff Bonnett took over) and it was arranged that he would have Head of Department status.

> "I have already agreed with my Railway Directors that they would not exercise their normal respective executive Departmental responsibilities for the operation and maintenance of the railway when it is built, and therefore would not have their normal responsibility for its planning and design. In these novel circumstances it is therefore essential to define clearly what responsibilities would *not* be borne by LT Departments, and who instead would bear them. I conclude with the following recommendations — a Light Rail Board should be established; the PM should report to the MD (Railways); the PM should be empowered to employ consultants and let contracts (but not for purchase of land), subject to approval by myself at certain levels; the PM should seek support from LT departments, as he sees fit, it being understood that he may prefer to use the services of external consultants — this, however, would not reduce his overall total responsibility for all aspects of the project, (a corollary being that other HoDs are under no obligation to assist if they do not wish to

do so); the Bus Business be asked to produce detailed proposals following up their earlier suggestions of an improved bus system for Docklands."

The system we developed was 'modern' in that it was *driverless*, staffed only by Train Captains, and no station staff. None of this would have been likely if it had been developed within London Underground. Furthermore we also opened without union involvement, although this changed later.

In the event the 'package deal' did not work as well as it should have because, urged on by the Heseltine philosophy, we handed over too much freedom to the GEC consortium. The chickens came home to roost with a vengeance when the commissioning, opening and early operation of the railway were profoundly complicated by the change of scope of the project required by the extension to Bank Station. Years later, the whole concept of 'systems integration' is a rapidly developing art, but was still in its infancy at the time. Furthermore, the concept of an intelligent client dealing with a consortium was not well developed. Yet, in retrospect, I have no doubt that the risks that were taken, particularly with the extension, described below, proved enormously worthwhile.

Lesson — Adding capacity to a railway during construction is complex.

## 8.3 Extension to the City of London

The project proceeded until, one August day in 1985, Sandy Csobaji — a Hungarian–American — came to my office. "Dr Ridley", he said, "I represent G Ware Travelstead, and we are intending to put a major new development in the Isle of Dogs. We want to put a major office building on top of your railway, and what is more we would like to pay to have you extend the railway westwards beyond Tower Hill, underground to Bank Station in the City." The first formal meeting with Travelstead took place early in the morning of 14 September. He firmly stated, "We want you to put a Bill through Parliament for this to happen".

### 8.3.1 The Role of the Barely Remembered G Ware Travelstead

The crucial role of 'Gee Whizz' Travelstead in the development of Docklands, long before Paul Reichman had ever been heard of, is rarely recognised.

Acting on behalf of two banks — Morgan Stanley and Credit Suisse First Boston — he had realised that London would not be capable of responding, within the environs of the City, to the 'Big Bang' revolution about to take place in banking. This was to require vast amounts of open, unencumbered, space and Travelstead decided that the Isle of Dogs was the place to put it. He fervently believed in the Docklands development and was a remarkable enthusiast.

Without doubt he was the greatest promoter I have ever met. He went everywhere from trade unions to Prince Charles, and talked to everyone to sell his idea. I remember one small example of his style — hiring the Orient Express to take a trainload of black-tied guests for an evening at Leeds Castle. However, Ware needed a physical link with the existing City of London but, additionally and importantly, a psychological link with the Bank of England, adjacent to Bank Station on the Underground.

The proposed extension had several impacts. First, it is costly to build underground. Secondly, the Health and Safety requirements for a railway are much more stringent underground. In addition, the capacity requirements (with double the number of passengers to be provided for) were very significantly increased. Thus, the signalling and control system for a driverless train that had never been developed before in the UK, had much more challenging requirements imposed on it, when it was half built but still only half developed.

My response to Sandy Csobaji, who was to become a professional friend and collaborator, was to say when I first met him, "Look, it's August, don't you understand that in this country we deposit Bills in Parliament to develop railways only once a year in November, and that's our only chance. I read about you Americans in the Times in May. It's now August and there's no time left to do what you want me to do. Where the hell have you been?"

Csobaji then finessed me to a greater extent than I have ever been finessed, before or since. "Dr Ridley", he said, " I do apologise, but you must understand that when Ware and I came into town, we were told that if you want to make any progress in London there were two organisations to avoid, because they are so terribly bureaucratic and old-fashioned — the Port of London Authority and London Transport".

I thought about throwing him out of my office and telling him to come back when he had a proposal for us to deposit a Bill 15 months hence, instead of three. But he had cleverly challenged me and, three months later in November 1985, we deposited a Bill to extend the DLR to Bank.

Lesson — To 'make things happen' it is sometimes necessary to grasp opportunities as they occur.

Of course, the day after the Bill was deposited we sat down and said to ourselves — what exactly do we need to do now that we have time to think about it properly? It is perfectly possible to propose amendments to your own Bill, if that seems desirable. We quickly realised that we had put the station at bank in the wrong place, under the Mansion House which had a very dodgy foundation. Fortunately, moving the station under the Northern Line could be achieved without needing extra powers, also giving a much better interchange with the District and Circle Line.

This was a ridiculous way to proceed with the project, but it was the only way open if progress was to be made. Ideally, our client body — LDDC — should have come to us and said we have asked you to do one thing, now we want you to do something different because our plans for developing Docklands have changed radically as the result of the arrival of an exciting new proposal from a group of American developers.

I have often wondered what would have happened if we had proceeded properly and taken the time to consider and plan carefully. If I had thrown Sandy out, would Travelstead have stayed around for an extra 12 months? Would the DLR have never happened? Would there have been a Canary Wharf? I believe it would have been pursued, but the delay would have been very unhelpful to the development programme. I have no doubt that I made the right decision to proceed with the ridiculous timescale for depositing the Bill.

A new extended contract with GEC-Mowlem now had to be negotiated. In particular we doubled the length of platforms and the trains. Travelstead's original 'offer' was £30 million but that came nowhere near the total cost of the extension. LDDC and the government, which was desperately keen to see the LDDC succeed, fairly readily agreed to pay the difference.

In the event Travelstead's backers — Morgan Stanley and CSFB — decided that the risk of their development was too great. But by that stage LDDC, having changed its plans, was totally committed. Their original plans for low-rise housing and industrial building had been completely transformed. Meanwhile, the City, led by Michael Cassidy the Planning Committee Chairman, which had previously had very strict planning controls preventing massive development, had changed their policy and planning regulations to fight back against the Canary Wharf and associated developments. They were desperate not to lose their prime tenants. Ware, for his part, was touring the City seeking to persuade banks and others to move. Not surprisingly it was the foreign bankers who showed most signs of responding. But it was not enough for Ware's backers.

## 8.4 Opening of the DLR

When Keith Bright wrote inviting the Queen to open the DLR in the second half of July1987 he received the standard 'too soon to say' reply; then in February from Buckingham Palace (Sir William Heseltine to LT Chairman, 2 February 1987), an acceptance for 30 July. It also said that, thereafter Sir Robert Fellowes, Deputy Private Secretary, would be the contact.

The completion of the DLR was always going to be touch and go, and there was great pressure to delay the opening in general and the Royal Opening in particular. GEC had various systems problems and, it being a package contract, we were one step removed from tackling the problems. On 26 May the Department of Transport wrote, (DoT to LT Chairman, 26 May 1987)

> "I have recently discussed with Tony Ridley and Cliff Bonnett certain problems relating to the handover of the DLR from GEC and the prospect that the planned date for the Royal Opening might be at risk. An important activity, currently on the critical path affecting the decision, is the technical audit being carried out by Mr Lawrence for DLR, whose start has been postponed because of delay in the delivery of GEC documentation. However, it appears that we need to have established no later than 19 June whether the opening can go ahead as planned, if the constraints imposed by the Queen's timetable and the need to issue invitations for the Opening are both to be satisfied. In the light of this, I should be grateful to know whether you believe the technical audit and other activity concerned with the commissioning of the railway can be completed, so that a firm decision on the Opening can be taken then."

On 15 June, after discussions with Bonnett and myself, Bright replied, (LT Chairman to DoT, 15 June 1987)

"As you know, the contractor has not finished all his commissioning work by the contract completion date set for 26 April. However, most had been carried out by 14 May and DLR took control on 18 May. Since that date driver training in both automatic and manual modes of operation has been completed and trial running has begun. The contractor is continuing to deal with final commissioning modifications as they arise, including all matters raised by the Railway Inspectorate. Wheel re-profiling is proceeding according to programme and the running of trains is bringing improvements to rail surface adhesion."

"It is planned to open the service at slightly lower speeds and with less frequency than when running at full capacity, gradually increasing as improvements are obtained. I am confident that there is no reason at this stage to doubt our ability to open a safe and efficient service on 31 July and that the planned date for the Royal Opening should therefore stand as 30 July."

Goldman replied that the Palace had been informed accordingly. Of course, embarrassment for the Queen had to be avoided, but there was another problem. The date for passenger opening was unclear but there was concern that, if the Royal Opening was delayed, the drive for passenger opening would slip back — perhaps for some months.

### 8.4.1 To Delay or Go Ahead with the Royal Opening

In the event the opening of passenger service had to be delayed. On 2 July I wrote to GEC (Ridley T to A Weinstock, 27 July 1987),

"Dear Lord Weinstock, I am faxing to you this brief statement on the reason for postponing the opening of passenger service on 31 July of the DLR. I might say that Bob Davidson and others representing you at my meeting with them yesterday, Sunday, concurred with our decision."

"The original trial running period for the DLR was 8 weeks. This is the period during which a properly operating railway is handed over to the operators to make sure that they are fully trained and experienced by the time the railway opens to passenger service. This is especially important in the case of a railway with novel technical and system features. All of us (yourselves and ourselves included) have been determined to see the railway open on 31 July, but it is in neither your interest nor ours to open for passenger service unless we can guarantee a sufficiently regular and reliable service."

"After a series of unfortunate incidents and occurrences we decided on Friday, 17 July that, at an absolute minimum, the railway must have nine full working days of satisfactory service — satisfactory under the assumption that there were passengers riding. Monday, 20 and Tuesday, 21 July were hopeful, though not wholly satisfactory, but the performance has deteriorated markedly since then, system wide and other failures persisted until and including Sunday, 26 July. Thus we have no confidence that we can guarantee a satisfactorily reliable service to the public and have decided as follows — after a week during which there is an acceptable minimum of system-wide and other serious failures, we will require an absolute minimum of two weeks of trial running before announcing the date of opening to the public."

On the same day I wrote to Bob Davidson, (Ridley T to R Davidson, 27 July 1987) confirming the criteria which would determine when the two week period of trial running could begin, which we had agreed at a 'crisis meeting' on the previous day, Sunday. Arnold replied (Weinstock A to T Ridley, 30 July 1987) 'Dear Tony, We are naturally disappointed that it has not been thought right to start the full passenger service tomorrow as scheduled, but we must agree that it would be imprudent to do so unless there is confidence that everything will run smoothly and continuously. I am sorry I was prevented from attending the official opening today. I hope it all went well'. I suspect he knew it had when he wrote it. The railway opened for passengers on 31 August. It was not without further problems, but anyone visiting Docklands today can only marvel at what grew out of the 'Toy Train' and what the DLR has enabled.

When David Hardy stood down as Chairman of DLR I took on that role. In my Chairman's statement for the year to 31 March 1988, (Ridley T — Chairman's Statement, 1988), I said,

"It is with considerable pleasure that I introduce this report on the progress and operation of the DLR during the financial year up to the end of March 1988. The £77 million investment was funded equally by the LDDC and

**PHOTO 8.1.   DLR Passenger Service was delayed but the Royal Opening went ahead On Time.**

*Source*: By kind permission of EW Scripps Co.

LRT. Their co-operation and the appreciated efforts of our staff have initiated a local amenity that we believe will be invaluable in the social and commercial regeneration of Docklands. There are many aspects of the railway which are new. New technology has been applied to providing transport for inner cities. At the same time, well tried and tested principles of railway operation have been retained with the objective of blending the best practices of the past with the most up to date technology. DLR has already shown the ability of light rail to provide a system tailor-made for particular circumstances yet with the flexibility for extension and enhancement as demand changes."

"The railway has attracted a great deal of interest both from local users and worldwide. Its modern image is attractive and the use of staff to provide helpful and personal contact with the customers, rather than being isolated in cabs has been much appreciated. The perceived reliability of the system was not outstanding when operation commenced and there were more delays than were expected. A great deal of attention has been given to this resulting in considerable improvement. However, the Board looks to see continued improvements in performance in the coming year."

"Passenger security has been good since the railway opened with no reported incidents of passenger assault either on a train or at a station. This is due largely to the high level of supervision by DLR staff, including the use of CCTV on all platforms, and also reflects the excellent liaison with the two police forces involved. DLR has also actively marketed its services, and has attracted media attention. Coverage has generally been favourable, and the public have responded well to the various marketing initiatives, which include a staffed information point and exhibition at Island Gardens station. The Company's policy of providing full wheelchair access to the system by means of lifts, minimal gaps between platforms and train floors and dedicated wheelchair spaces on trains, has been well received. This facility is also popular with, and well used by, those with pushchairs and shopping trolleys."

"DLR was formally opened by Her Majesty the Queen on 30 July 1987 and commenced public service on August Bank Holiday, 31 August 1987. In the first months of operation approximately 16,000 passengers a day were being carried on typical weekdays. This has shown a steady increase to a daily total of about 20,000 passengers travelling on the railway by the end of the financial year. During the seven months from 31 August 1987 to 31 March 1988 the railway carried approaching four million passengers. This is most encouraging and shows a growing confidence in the new transport facility."

Finally, before leaving LT in November 1988, I gave a paper on financing urban transport (Ridley T, 1988), using the DLR as an example. In it I quoted one of my favourite stories about the DLR, previously told by my colleague, John Willis,

"Less than five years ago a public meeting was arranged to discuss the new idea for transport in Docklands — a light railway. Several thousand leaflets had been distributed, posters displayed and notices published in the local press. On the night, about eight members of the public turned up, outnumbered by officials two to one. The concept of light rail was explained and how it could dramatically improve accessibility to the area and exciting impressions of things to come were produced. But, of course, none of the eight believed a word of it. They had heard that someone from London Transport was talking about transport in Docklands and all they were really interested in was why there was a 25 minute gap in the 277 bus service a week last Thursday — and it was raining!"

The rest is history, wrote Willis, and I now jump forward 16 years to my involvement in the Olympic Bid 2004–2005.

## 8.5  It Was the 'Toy Train' Wot Won It

London, 6 July 2005. Jacques Rogge, the President of the International Olympic Committee, by a large television screen in Trafalgar Square. "And the winner of the 2012 Olympic Games is … (pause) … the city of London." Pandemonium.

I have borrowed the title of this from the Sun, 'It Was the Sun Wot Won It', when it claimed that its support in the 1992 election gave John Major an unlikely victory. The Olympic success owed much to Tony Blair, to Ken Livingstone, to Seb Coe and many others, but it is interesting to speculate whether it would have been even vaguely possible without the coming into being of the DLR — the 'Toy Train'.

The development of the DLR was the very antithesis of everything I believed early in my career as a transport planner — carefully thought out strategy, accompanied by well worked out plans and implementation processes. But, when you become involved in making things happen, planning sometimes has to give way to, or at least make room for, the art of the possible, snatching opportunities as they pass fleetingly by.

I believe that a virtuous circle was set in motion. Michael Heseltine begat the Docklands development, which begat the DLR, which begat Canary Wharf, which begat the DLR Extension, underground to Bank. The DLR Extension and Canary Wharf begat the Jubilee Extension, which begat the Channel Tunnel Rail Link through Stratford. And so on to the successful 2012 Olympic Bid.

Lesson — Creating new transport links creates opportunities for growth.

When work began on the DLR we were regarded as mad for developing a White Elephant. With the arrival of Travelstead we were castigated for having built a totally inadequate system. I had considerable pleasure in arguing that there must have been a two-minute window in history when it was 'just right.'

We shall never know whether the linkages (the begats) were dependent on each previous step, as I describe, but I would contend that it is at least a tenable hypothesis. Indeed, I had written a piece for *the Times* (10 October 1991) when joining academia, arguing that 'the high-speed line via Stratford could transform our transport policy,' Furthermore, I said, 'a decision which unites Gerry Bowden, Tory MP for Dulwich, who does not want the link in South London and Tony Banks, Labour MP for Newham North-West, who does want it in East London, cannot be all bad.' Transport is politics.

It may also be possible to contend that the rebirth of Gateshead and Newcastle was equally dependent on the creation of the Tyne and Wear Metro. Today, no-one doubts that there is a link between the role of new transport links and opportunities for growth.

## 8.6 Transport — Prerequisite to Winning the Games

In January 2004, an email arrived from a certain Mike Power. 'Tony. By way of introduction, I am the COO of London 2012, the Company charged with the mission to bring the Olympic Games to London in 2012. I am charged with putting the technical part of the bid together, so that Barbara Cassani and Keith Mills have something to sell to the government, press, country and, importantly, the IOC. You and I have met on a couple of occasions, the first being in the early 1970s when I was working in Operational Research at P&G and you came to speak to the North Eastern Group of the OR Society (in the Three Tuns at Durham if my memory serves me correctly)'.

The IOC was the International Olympic Committee, based in Lausanne. Barbara was the Chairman of London 2012 and Keith the CEO. Proctor and Gamble was the American conglomerate whose UK headquarters was in Newcastle.

Power explained, 'One of the most contentious parts of the bid is Transport and, though we are getting excellent service from TFL, I feel the need for some independent advice. Put simply, TfL is not impartial to some of the outcomes and there are times when I feel we are getting squeezed between the DfT and TfL agendas and politics. Hence, someone with an independent frame of mind who could provide us with ad hoc advice would be a real asset to the bid. Likely we will get some help from KBR (a firm of consultants), who were heavily involved with the Sydney games and are now involved with Athens, but their expertise is 'Olympic Games travel' and their lack of knowledge of London is going to be a handicap. To fill the gap, I am looking for someone who has the knowledge, skill and expertise re transport in London to act as an independent part-time advisor. Hence the note and a request whether you would have an interest in doing this or whether you can point me at someone who might.'

After two conversations I agreed to join as Senior Transport Advisor — working three days a month, except at the time of the visit of the IOC Evaluation Commission in early 2005 to hear our bid proposal. I undertook to work as long as it took at that time, but I was determined to avoid becoming a part-time executive. I had asked Mike, "Why me?" "Well, you know transport." "Yes." "You know London." "Yes." "Thirdly, the IOC's transport advisor is Professor Philip Bovy, of the University of Lausanne. We thought that if they had a professor, we'd better have one too!". "OK, what do you want me to start with?" "Go and get a grip of that so and so Sumner". "What, Hugh? He used to work for me at LUL. What's his problem?". "He's impossible to work with." I went to see him. "Hugh, you've really upset a lot of people." He explained what he had achieved during the previous six months, with a small team within Transport for London. Transport was probably the most significant challenge to the London bid. It immediately became clear to me that without Hugh and his team, the whole bid would already have been 'dead in the water'.

## 8.7 The Job

I soon realised that my principal job was not transport, nor London, but to encourage/help the TfL team and the Olympic Bid team work together effectively. Thus, once again, I confirmed that the most important aspects of my career during the latter 20 years were what I had learned about interpersonal relationships and group behaviour.

Lesson — My Olympic role did not need so much planning, engineering, or even political skills, but needed interpersonal skills.

The Chairperson of the Bid team was an American lady, Barbara Cassini. No doubt an excellent businesswoman, she did not last long. Keith Mills, who had made millions out of Airmiles, was a 'brilliant marketer', I was told. He was the CEO but he was also worth his weight in gold when it came to keeping tabs on the likely voting preferences of delegates when the decision came to be made. Power reported to Mills.

I began to understand that I was a 20<sup>th</sup> century person when I asked for an organisation chart. It took many weeks to let me have one. I was gob-smacked when someone tipped me off that the problem was telling some of the managers that they were not, in fact, reporting to the senior level they had assumed!

But there was a more fundamental problem. Most of the Bid team, who were intelligent and very hard working, seemed to be convinced that although the competition was winnable, it could not be done because of the terrible transport offering that would kill the Bid. Some of Cassani's closest advisers whispered this into her ear. So there was a big selling job for transport, ultimately to the IOC, but first to the Bid team itself.

When Seb Coe arrived I was not necessarily convinced that he was the man to fix things. I could not have been more wrong. Like all of my generation I well knew of his athletic super-achievements, and it was very useful in his dealings with the IOC that the UK bid was led by a world-class athlete. But, when he came, I regarded him as former, not very impressive, Tory MP who did workouts with the one-time Leader of the Opposition, William Hague. Perhaps I am a rather poor judge of people at a distance. He turned out to be an inspiring leader, who was one of the main reasons that we won. At a stroke, he also changed the team's view of Transport. After asking for a briefing he began to say — repeatedly — to the team, the press and to the government and the IOC, "Transport is our strong suit!"

The preparations for the visit of the Evaluation Commission were meticulous. It had originally been decided that the Transport presenters would be Ridley, Wilben Short, who was full time with the bid team, and Sumner — in that order. We had gone a long way when someone decided that the Minister of State in DfT, should be part of the presentation.

The scripts were developed, rewritten, topped and tailed, tweaked, lengthened, and shortened. We were 'coached,' at great length. I was sceptical at first, but began to see the point. By simplifying the layout of the notes, and adding highlighting, it was possible to take one sentence, or a short paragraph, at a time — thus allowing more direct delivery to the audience, and eventually to the Evaluation Commission, in a way that allowed eye contact, considered to be extremely important by our coach, an American lady.

We were coached for the Q&A session. Always start with the 'promise,' and then elucidate. We were invited to wear dark suits, with a white or pale blue shirt. A standard tie was handed out. We were ready, after numerous rehearsals, through which the presentations became increasingly slicker, though the Minister, not having been to any of the rehearsals and having rejected the need for a politician to be told how to speak 'by an American', demonstrated how far we had come when, at the final dress rehearsal, he shuffled his papers and read most of his notes head down.

The presentation to the Evaluation Commission was on the next day. The plan was that Seb Coe would introduce me, and handle the Q&A session afterwards. Seb was excellent, in command and authoritative. We were delighted that he would take the role. We were 'on' at 2.50 pm — lined up, stood in line according to our seating plan, and marched in. During the rehearsals we developed the arrangement whereby our Theme leader would sit next to Seb so that he could indicate smoothly to him who should answer each of the questions. It was also important that he and Hugh should be adjacent, as Hugh was likely to handle the greatest number of the questions.

Coe was to commence with a build-up to Ridley the Academic, but with the Hong Kong system to his credit. Then I was to establish my bono fides and introduce the team — and also to emphasise the tremendous amount of work that had been put into Transport. That was the plan, but there was a hiccough.

Many late changes were made to the script, and there were late changes to the team and during the Q&A session we had to 'wing it'. However, we seem to have satisfied the examiners and won the Bid. As everyone knows the London Games 2012 were a major success with transport being outstanding.

My final small involvement with London's Olympic programme was to chair an all-day conference at One Great George Street, 'How London 2012 Will Change UK Transport', being transport's contribution to the ODA's wider Learning Legacy exercise (ODA, 2012a). The conference was sponsored, *inter alia*, by ICE, IET and the APM. Transport's contribution was summed up in the Executive Summary of 'Delivering Transport for the London 2012 Games' (ODA, 2012b), while the programme (ODA, 2012c) contained four sessions, with eight presentations and 12 workshops, in five rooms. Both the opening and closing addresses were given by me to some 300 attendees (Ridley T, 2012).

# Bibliography

Bridgeman J to Chairman (7 October 1982).
Bridgeman J to J Swaffield (24 November 1982).
Chairman to DoT (15 June 1987).
DoT press release (7 Oct 1982).
DoT to Chairman (26 May 1987).
Heseltine W to Chairman (2 February 1987).
ODA (2012a), Learning Legacy, Special Supplement.
ODA (2012b), Delivering transport for the London 2012 Games, Executive Summary.
ODA (2012c), How London 2012 will change UK transport, Conference Programme.
Ridley T (1983), DLR project, Memo to LT Executive Committee, 15 March.
Ridley T, DLR Chairman's Statement, 1988.
Ridley T (1988), The Docklands experience, the move to private funding, In *Financing urban transport: Sources and techniques*, IMTA Meeting, London, 13 April.
Ridley T (2012), How London 2012 will change UK Transport, Opening and Closing Addresses, Conference, ICE, 26 October.


# Press

Ridley T (1991), *Times*, 10 October.


# Personal Correspondence

Ridley T to R Davidson (27 July 1987).
Ridley T to A Weinstock (27 July 1987).
Weinstock A to T Ridley (30 July 1987).

# 9

# Eurotunnel 1987–1990

**Many decisions about this internationally significant project were taken before an owner/client had been identified. How did this affect the project? Could there have been any other outcome than the client and contractor in constant dispute with each other. Was this the best way to set up a project to be successful?**

Some lessons

- Be very careful about the terms of reference of a job before you take it.
- Feel free to pull the wool over your adversary's eyes, but never pull the wool over your own eyes.
- By mutual agreement is a good way to leave a project.

## 9.1 How It Began

After the trauma and complexity of the King's Cross fire and my resignation, there was some uncertainty as to what I might do next. However, as Richard Hope had forecasted, it did not take long for me to join ET after I left LUL. This presented the new experience, of joining a project mid-stream which was already embroiled in publicly-aired disputes between contractors and client.

The creation of the Channel Tunnel was, in many ways, a triumph of engineering. Indeed, in a narrow sense, it can be regarded as 'the project of the (20th) century'. Sir Alistair Morton was seen as the person without whom it would never have been completed. Yet, when I later became an academic,

I used to tell my students that the HK MTR was an example of everything right about developing a Civil Engineering project, but that the Chunnel was an example of much that was wrong. How can this be?

This chapter is not 'the story' of the project, merely my experience. Happily, the early days of the project which completed the Chunnel, 200 years after it was first mooted, are recorded in Sir Nicholas Henderson's reflections (1987). A retired Foreign Office mandarin, he became the first Chairman of the Channel Tunnel Group, the five British contractors and two banks who subsequently merged with France Manche to create the first version of Eurotunnel. Henderson's CTG board consisted of some of the big names of the past — Don Holland of Balfour Beatty; John Reeve, then Tyrrel Wyatt of Costain; Alan Osborne of Tarmac; Frank Gibb of Taylor Woodrow; Andrew McDowall, then Chris Chetwood of Wimpey; together with National Westminster and Midland Banks. Tony Gueterbock (now Lord Berkeley) and Alistair Dick, of my generation, are also mentioned. Coincidentally, it had been McDowall and Gueterbock who had approached me, several years earlier, to interest me in the project.

In an engaging account of how the competition for the concession was won, Henderson concludes, 'We will traverse many difficult times before we go through the tunnel by train or shuttle. I am as sure of the recurrence of frequent crises as I am of the ultimate successful completion of the project'. True, if substantially understated.

Another good description of the project is by Fetherston (1997), who includes an interview with me. He quotes me, 'The tunnel isn't the greatest Civil Engineering project in the world. It's the greatest transportation system, which happens to run through three big and expensive holes under the Channel.'

My first significant contact with the project occurred early in my time as MD of the Underground. The telephone rang and a voice said, "I am Alistair Morton and I have heard a lot about you. I would like to have lunch with you." I said yes. Like any of us, he had his plusses and minuses, but he was a very dynamic and lively individual. He said, "we've got real problems with our project and we would like you to come and talk to some people here". He did not say, "I am offering you a job." In fact he was strangely opaque about what he had in mind.

I also met Pierre Durand-Rival, who came from the French steel industry and was the director of the project and operations. He was the technical man in the broadest sense. There were financiers, lawyers, two Chairmen and the like, but Durand-Rival was central to the whole task. I met Morton again the following day, when he made it clear he would like me to join the project. It was obviously an interesting idea but I now, in effect, had an entirely new job with the Underground. When I joined the ridership had been declining for 30 years. Then, for a series of reasons, ridership rose 60 percent in five years. I was at the beginning of a brand new phase which I did not want to leave. I replied that I already had a perfectly good job and wasn't keen on Durand-Rival. So he asked, "are you prepared to come on the Board?" "Certainly" I said.

There were two Chairmen of the Board of ET — Morton, who was a South African, and Andre Bernard, the most senior Frenchman ever to have worked for Anglo-Dutch Shell. Because ET was a bi-national project they had to have two chairmen to take care of their respective constituencies. Although Morton was a South African he had lived and worked in Britain for a long time. After Oxford he studied at MIT and worked for Anglo-American. He was an early 'young Turk' with the Industrial Reorganisation Corporation, and later became MD of the British National Oil Corporation, under the chairmanship of Lord Kearton. Nearly everyone who knew him had some adverse comments to make about his style of doing business which could, of course, have indicated that his successes had rankled.

I joined the Board as a non-executive on the same day as the illustrious 'Black Bob' Scholey, the Chairman of British Steel. I thought, 'well I've arrived in the big world with all these big names.' We would meet as a Board and give technical advice about the project. It was clear that things were not going well, in particular the relationships. Little did I realise at the time that the project was set up backwards, the contractors having been appointed before the Owner was created and, while it might be completed, it was never going to be successful — at least by my definition.

### 9.1.1 Morton in Action

I had been on the Board for more than a year, attending monthly meetings alternately in London and Paris, when one evening at a social event at the Festival Hall, Alistair was present. He could be very engaging. "Come over here I want to talk to you" he said, "I want you to join the project". "Alistair", I said,

"I've told you I've got a job, I am not coming, I'm already a Board member. What are you talking about? I don't want to come and work for Eurotunnel". "Oh I don't want you to come and work for Eurotunnel", he said, "I want you to work for TML, the contractors". "But you're the Co-Chairman of Euro-tunnel, how can you do that?" I said. "I'll tell Frank Gibb" he replied.

Gibb, who was the Chairman of Taylor Woodrow, was then also the lead for the five British contractors. "I'll tell Frank Gibb he needs some proper people working for him … go and talk to him". I did ring and said, "Alistair says he's been talking to you and he wants me to discuss the possibility of coming to join TML". Then I sat down with him. It wasn't clear what they wanted me to do. However, my usual sense is that it is always useful to talk and, if the client said Ridley is someone you ought to have on your team, it was worthwhile for the contractors to talk at least. If there was a potential role, it would be the contractors' appointment, with Frank in charge. However nothing quickly came of this as they manoeuvred backwards and forwards, between themselves and the French.

As often happens, events take over (in this case the King's Cross fire). On 9 November I rang Morton saying, "You have been pressing me to join ET, or TML. You may in fact want me to resign from the ET Board because I am going to resign from London Transport and the Underground tomorrow". He could, and did, make game changing decisions with lightning speed. "Give me an hour and a half", he said. At the end of that time he rang back. "We want you to join the project", he said.

Hence, four days before Christmas 1988, though there was still unfin-ished business with Frank Gibb from whom I had not heard for some time, Morton called. "I want you to come and have breakfast with Andre and me in the Goring hotel at 8 o'clock tomorrow morning".

Straight down to business. "We want you to take the job of Eurotunnel Project Director. Durand-Rival is moving to a non-executive role and Alain Bertrand will be Operations Director, with responsibility for safety". He named a salary and said I should start in mid-January when a new agree-ment was expected to be signed with TML. I was obviously available, as I was now 'between jobs', but I said, "what about the conversations I am having with Frank at your instigation?" "Oh, we can't waste our time waiting for him, he has dragged along far too long". It was clear that Benard was at one with this.

### 9.1.2 A Serious Mistake

At this stage I made what I regard as a serious mistake. Saying yes was not in itself a mistake, but tactically I mishandled it. The offer was clearly a professional and public vote of confidence after my King's Cross experience, and I would be gainfully employed again after only just over a month's break. What I should have done, however, was first to go back and talk to Frank, if only to respect our conversations. Additionally, I should have pressed for clarity about my terms of reference relative to the two chairmen, if I joined them. I did not. Perhaps I was still 'injured' by my King's Cross experience and not thinking clearly. The deal was agreed there and then, rather too hastily. The two Chairmen were going off for their respective vacations that very day. Lesson — Be very careful about the terms of reference of a job before you take it.

The next morning at 9 o'clock, another phone call. It was Gibb. He had a friendly but very hesitant style. "Tony", he said, "I'm very sorry to have taken so long. I have talked to all 10 TML contractors, and I have the authority to offer you the post of Deputy Chairman of TML" i.e. number two to Phillip Essig. Philip was a French friend whom I first knew when he ran RATP in Paris. "Bloody hell, Frank", I said, "just yesterday I agreed to join ET as Project Director". "Oh dear, what can we do?" "You'll have to talk to the Chairmen, but Alistair is in the Okavango swamp in Central Africa. Andre is at his place on the Mediterranean". We finished our call with him saying he would call Benard. Gibb called back later. Benard had said 'No, Ridley is our man'.

I can reflect endlessly on how it would have turned out if I had joined TML. It would have been very tough, but I do believe I would have found it much easier to negotiate across the table with Morton than to work with and for him. We shall never know. Thus, I joined the Board as a director at its meeting on 16 January 1989.

## 9.2 Starting as MD

My arrival in a full-time role coincided with the signing of a Joint Accord between ET and TML. The Accord, signed on 16 January, said, 'Following discussions in September 1988 between ET and the Instructing Banks leading up to ET's first drawdown under its Credit Agreement, ET and TML subscribed to a Memorandum (which reads) — ET and TML agree to exercise their best endeavours to improve their working relations, increase the efficiency of their respective organisations, settle as far as practicable the outstanding issues between them and find positive ways forward to complete the work within time and cost frameworks to be agreed.'

The *Financial Times* (13 January 1989) reported that Ridley had been appointed joint MD of Eurotunnel. Morton said that I was "probably the most experienced mass transit construction engineer in the world" and that he had no qualms about the public relations impact of appointing me. "The most informed observers of what happened at King's Cross saw a great many plusses in Tony Ridley's activities at London Underground. 31 people were killed on an escalator that he did not build, of a type which he was in the process of removing. That escalator might have been located in a department store, but he was the boss and the buck stopped there." Morton also announced the appointment of Alan Bertrand as joint MD in charge of developing the design and development of the railway system which would operate through the tunnel. The article went on, 'It also emerged yesterday that Philippe Essig, a Former Chairman of SNCF has been appointed Chairman of TML, a job previously held by one of the ten members of the construction consortium.

*New Civil Engineer* (NCE) (6 April 1989) headlined 'Chunnel finds a clearer path.' Four months earlier NCE had said that the relationship between client and contractor was based on mistrust, suspicion and doubts about the other's motives and abilities. The resulting friction was bound to lead to serious problems with the job itself. It went on to note the fact that the appointments of Ridley and Essig had been welcomed by the other side, TML and ET respectively.

### 9.2.1 Rebuilding Relationships

When Ty Byrd interviewed Essig and me for a cover story (*NCE*, 18 May 1989), 'Team spirit lifts Chunnel outlook', we appeared on the front cover, he with a glass of (English) beer and I reciprocating with (French) wine.

The article said, 'Relations at a senior level between the two organisations had become very bitter. There was an urgent need to rethink both

**PHOTO 9.1  Anglo–French Accord.**
*Source*: By kind permission of Emap.

management structures, reach accord on various design and construction matters, and proceed in a more united manner.'

"The two appointments were deemed to be complementary although on opposite sides of the fence. Each man was considered right for his particular job: the fact that they knew each other and got on well, and that ET/TML relations might benefit as a consequence, was considered a bonus. ... This has helped the rapid generation of a new spirit of cooperation between the higher echelons of ET and TML. Both men took up their positions earlier this year and their first action was to tour together the Chunnel sites at Sangatte and Folkestone. According to Essig the high profile joint visits made a profound impact."

"Relations had been very poor although the reasons for this are not entirely clear", Ridley says. Essig cites the fact that the project was let as design and build but is so huge that it cannot be worked in a conventional way, involving ET in the process much more than is usual with a client. He believed this could have caused anxiety and increased antagonism within TML. Less pragmatically, Ridley feels that having nurtured ET, the founding companies of TML psychologically felt that they had a cuckoo in the nest. They were resentful when the client had to achieve, then demonstrate, its independence. He also makes the point about the depth of feeling within ET that TML saw the tunnel merely as a construction contract, not as an on-going transport system.

"Ridley and Essig's remit to change attitudes and practices was a substantial part of the Joint Accord signed on 16 January. After visiting both construction sites the two met with a dozen key senior staff members to talk about formal relationships, clarify duties and generally discuss how the Chunnel should be managed. What emerged most clearly from the two-day session was the need for closer liaison between the two organisations."

"Better communication would promote a virtuous circle of better relations, greater mutual confidence, more effective project control and higher performance. Ridley now chairs a monthly ET/TML executive committee, which Essig attends, to analyse progress and cost and allied issues. They chair between them a transportation committee, set initially to meet monthly but now convening more frequently as the myriad of interfaces between civil engineering and the railway's electrical and mechanical systems looms ever larger."

### 9.2.2 Nichols and Eurotunnel

At one stage, in 1989, in my frustration with the appalling relationship between client and contractors, I had decided to try to introduce some of the Hong Kong ethos (win together or lose together) into the ET project. I turned to Mike Nichols, who appeared in both the Hong Kong and Tube chapters, to carry out a review on my behalf. To this end he brought two of my Hong Kong MTR colleagues — Ron Mead, Project Director, and Michael Yeeles, Chief Accountant — and both then self-employed consultants, onto his team. On 5 April (Ridley T, 1989) I sent a memo to senior ET staff

"In the time that I have been with the ET project considerable progress has been achieved. I believe that relations with TML are much improved and that the structure of decision-making for the transport system is developing, with more joint working. Considerable effort has been put into the examination of the costs of the project and the rate of tunnel progress has improved. However the project programme is pressing hard upon us and not nearly enough has been accomplished on some critical strategic issues — such as overall project policy, organisation, cost structure and comprehensive programme. I want to make more rapid progress in tackling these issues so as to enhance project performance and improve the prospects for a successful completion of the project. Therefore I have appointed Nichols Associates to assist me in this task. They are consultants specialising in project management, with particular emphasis on strategy. They have considerable experience of large-scale railway projects and have carried out a number of assignments for me in the past."

My team did collaborate, but I probably overlooked the fact that Bechtel (ET's consultants) were never happy to have anyone move into what they regarded as 'their territory'.

The aim at the outset was that Nichols should propose short-term measures to improve project performance but, after an initial assessment, Alistair and I agreed on 13 April that more radical proposals should be considered. Nichols produced a draft report of their review on 2 June (Nichols Associates, 1989a), having interviewed five senior managers of ET, including myself, and eleven from TML, they had quickly concluded that the type of project was not unique.

"The granting by governments of concessions to build and subsequently operate public utilities and transportation systems, on a privately owned and commercially run basis, is by no means unusual. … A revival of interest in recent years in this type of venture has resulted in a number of concessions, and hence construction projects, of this type in various countries."

"Nevertheless the Channel Tunnel Concession embraces a number of exceptional features which greatly complicate the task of carrying the project through to a successful completion. The most significant of these are — international importance; scale; funding; consortium structure; and technical requirements. These features have combined to produce a high level of complexity in organisation, design, construction and operation which must be rationalised within the time and cost constraints of a commercially viable fixed link. In the light of these circumstances, any recommendations which seek to vary the interplay of responsibilities between the many parties involved in the project — especially now that almost half of the contract period has elapsed — can only be predicated on the basis that, unless the recommendations are adopted, the Fixed Link is most unlikely to be completed within acceptable time and cost limits."

On contractual relationships they went on,

"The specification for the Fixed Link, as embodied in the Construction Contract, necessarily left substantial areas of both requirement and detail to be subsequently developed and agreed. It is self-evident that a contract price can only be as fixed as the specification that fixes it."

"Failure to recognise this simple truth will inevitably lead to disputes. Increasingly, the Employer will claim that virtually every subsequent development of the specification and design was included in the Contract, and the Contractor will adopt the contrary position. By the time agreement has been reached in such circumstances, additional costs will have been incurred and time will have been lost. This has been the experience so far in relation to the Construction Contract, as evidenced by all the circumstances surrounding the requirement for the Joint Accord and the commonly held opinion that this agreement is unlikely to halt the history of successively increasing predictions of final total cost."

"The present project status gives serious cause for concern and that, if the present situation is allowed to continue, the prognosis is not good. In these circumstances our conclusion is that significant corrective action is urgently required. … We consider that the measures recommended in this report can and should be adopted provided they can be acted upon quickly and decisively. The next three to six months give the last window of opportunity for major changes to impact sufficiently in time to achieve a successful completion of the project."

I was hopeful that discussion with Morton might lead to consideration of radical reform. This might have been possible, but it became clear that Benard would not countenance any change. They were an interesting pair, completely different in character, but wisely committed to each other so that the contractors could not drive a wedge between them. The result was, however, that Benard could and did exercise a veto over any flexibility that the project might require, which Morton might have been willing to contemplate.

My initiative of using Nichols to chart a better way ahead was not successful, but I have included their Final Report recommendations (Nichols Associates, 1989b) in the Bibliography, for the possible benefit of my young engineer audience.

### 9.2.3 Jack Lemley

At about this time TML appointed the American Jack Lemley as Chief Executive, reporting to Essig. Ty Byrd had said,

> "New titles and job responsibilities have been or are about to be given to a number of existing key personalities, among them Andrew McDowall. McDowall had been TML's top man and was then designated deputy chairman on the arrival of Essig and may move again following the appointment of Lemley."
>
> "Ridley says he has never worked on such a complex project before and feels there must be areas where economies can be made. 'Quite apart from it being an Anglo-French project, we have an enormous client; a contractor with 10 major contractors behind it; four agent and 22 instructing banks; more than 200 other banks involved worldwide; a maître d' oeuvre; umpteen consultants and many more technical advisers. The complexity of relationships is enormous, the amount of reporting absolutely horrendous. We have to simplify the relationships and reduce the reporting back.'"
>
> "There seem to be no differences between Ridley and Essig on what constitute the priorities for getting the Chunnel up and running. Ridley lists them as, 'achieving consistently the tunnelling rates we have now demonstrated can be achieved; completing the design of a transportation system that ET can operate successfully until well into the 21st century; and working out practically the complex array of interfaces of the components of the project.'" (*NCE*, 18 May 1989)

With the arrival of Lemley my TML focus of attention switched from Essig, who was occupied with 'wrestling' with the two chairmen, to Jack. Fetherston (1997) describes him as fitting the image of a big-job engineer. 'He was tall and rugged, direct in speech, accustomed to command and entirely comfortable with it. He shook hands with a powerful but controlled grip and regulated his language in much the same way.'

Lemley and I spent much of the time between June and September trying to make the Joint Accord work properly and to find substantial cost savings, on both sides. The 'negotiations' were somewhat bizarre in that we were arrayed across a table, each with our support teams with us, but my team were not watching 'the enemy'. They were rather keeping an eye on me, on behalf of Benard in case I was tempted to 'sell the pass', or so it felt. "I couldn't do a deal with Lemley, because I couldn't go far enough on behalf of ET and he was not prepared to go far enough to meet us". (Fetherston, 1997). I went back to Benard and Morton, and quoted the line at the end of the Peace of Amiens during the Napoleonic era, 'Normal relations between England and France were resumed, they were at war.' And that's what happened between ET and TML which seemed impossible to modify — the war party took over again.

Now Benard intervened and took me out to lunch, for a friendly conversation but which, however, felt rather like a game of poker. He carefully explained that he and Morton had agreed from the outset that he would handle personnel matters with the French staff, and likewise Morton with the British. Thus, when Durand-Rival had to be stood down from his executive position it was he, Benard, who had acted. I was Morton's responsibility, but he wanted to discuss the state of play with the project and my role. He went on at some length at the end of which I said, in some puzzlement, "But Andre, if what you have said is true, then Alistair thinks that I am after his job." "Of course he does", he said, "why haven't you noticed?" I was absolutely flabbergasted. While we were talking about my relations with the contractors he also suggested that I had a problem of 'angelisme' — which I interpreted as naïve optimism.

At the beginning of 1990, having been MD of the project for a year, I had spent the time trying to persuade the parties to recognise the reality of what the project was going to cost. Self-evidently the problems were massive. The challenge was exacerbated by each side, particularly ET's two Chairmen preferring to row about the responsibility between the parties for cost increases rather than work together with the contractors to stop the project cost growing, the very antithesis of proper project management. Even worse, senior people in Eurotunnel seemed to be in denial about the likely cost of the rolling stock — certainly wool over the eyes.

Lesson — Feel free to pull the wool over your adversary's eyes, but never pull the wool over your own eyes.

At first relationships improved as I was able to work closely with Essig, and then to negotiate a way ahead with Lemley. At this point we were newcomers to the project. It soon became clear that these relationships were uneasy. In particular, Benard had little time for his compatriot Essig. Lemley was tough but a realist, and realism was not welcome.

### 9.2.4 End Game

The press were deeply engaged in following the 'difficulties' within the project, and were eager to report that war had broken out. Morton had failed in his determined attempt to forbid the contractors from talking to the press, while he used the press freely to insult and belittle them himself.

'TML wants top-level change at ET', said *Construction News* (18 January 1990), while *Telegraph* (25 January 1990) asked, 'Will Tony Ridley soon join the list of executives who have decided that there are easier ways of making a living than drilling the Channel Tunnel.

> 'In the claustrophobic world of tunnels and tunnelling, few have a higher reputation than Ridley, 57, after achievements with Hong Kong's cross-harbour tunnel (metro) and London Underground... Morton recently refused to be drawn on Ridley's future, and favours an adversarial relationship with the contractors drilling the tunnel. The contractors are not allowed to speak to the Press, but believe that Ridley, who understands their problems, is getting on too well with them for his own good.'

'Costain rips into Morton over ET statement', said the *Contract Journal* (25 January 1990) on the same day.

The *Independent* (26 January 1990) said that 'The Bank of England is supporting pressure from ET's bankers for Alistair Morton to surrender day-to-day management involvement in the Channel Tunnel project'. It went on to quote a letter, from Peter Costain, accusing ET of producing a financial report on the tunnel that was 'inaccurate, incomplete and calculated to mislead'. It further charged Morton with being 'provocative, tendentious and disingenuous.'

Two days later the *Sunday Times* Business Section wrote that Ridley might step down as Eurotunnel's managing director over what it described as "my increasingly invidious position". They said that the ten contractors that made up Transmanche Link were urging me to stand up to Alistair Morton, but that Alistair and other Eurotunnel executives believed that I was not cracking down hard enough on the contractors.

> 'Ridley's reputation as a tunnelling expert and railway operator was reinforced as managing director of Hong Kong Mass Transit Railway. At London Underground he devised the plans for new rail lines under the capital. There is a growing campaign by non-executive directors of ET, discreetly supported by the bank of England, to press Morton to appoint a chief executive to deal with the contractors. This would leave Morton free to liaise with the bankers. Ridley is not seen as a candidate for the chief executive slot. "Both he and Morton have powerful egos and I don't think it would work", said one source.'
>
> 'Problems between ET and TML have been endemic since Morton took a grip on the project, he is said to be constantly suspicious that the contractors are trying to pad out their prices. One ET source summed up the attitude to TML prevailing under Morton as "we've got to fight the bastards until the opening day." The contractors, for their part, are "very, very bitter about Morton" reports another source. This bitterness was expressed in a recently leaked letter to Morton from Peter Costain, chairman of the Costain group, one of the ten TML members. 'I cannot understand what you hope to gain by the disparaging remarks about the performance of UK construction', Costain wrote. 'Your comments do nothing to improve the relationship between us. They have an intensely demotivating effect on our employees. It does you no credit at all that, immediately following a difficult but successful negotiation, you should take such a provocative public line." (*Sunday Times*, 28 January 1990)

The *NCE* (1 February 1990) announced that 'ET may turn to US for top posts' and quoted contractors as saying, 'we wouldn't mind Ridley as

Chief Executive: he understands huge transport projects. But to admit that does the man no favours. Our approval would be the kiss of death for Ridley, as far as Morton is concerned'.

Three weeks later it said, under the heading 'French hardliners delay Chunnel deal', reported that funding had been a problem since September 1989 when TML and ET failed to agree project costs as they were obliged to do to ensure continuing draw-downs of money from the Channel Tunnel's syndicate of lending banks. Narrowing the gap between the two organisations on costs and other matters had been the subject of intense and sometimes acrimonious negotiations. Agreement had been near the previous week, and eventually depended on reorganisation of ET's top management in a manner of which TML approved.

> "The contractor wanted a buffer between itself and Morton. Then, last Thursday, ET announced changes to its management structure that included the apparent demotion of Morton from Co-Chairman to Deputy Chairman of ET but confirmed his continued 'hands on' control of construction with the added title Chief Executive. TML responded immediately and vigorously, refusing to sign the amending agreement and effectively putting a stop to all funding. The following day, with Governor Robin Leigh-Pemberton convening a meeting of ET headed up by Morton and TML personnel led by Tarmac's Neville Simms, Simms made it clear that a chief executive other than Morton was needed to control construction of the Chunnel for ET. In a three-hour meeting, agreement was achieved over the creation of a new role, that of project chief executive, to report through Morton to the ET board. The name of John Neerhout of Bechtel was raised as a suitable project engineer for the job." (*NCE* 22 Feb 1990)

The *Independent* (22 February 1990) by-lined a story, 'Dr Ridley speaks his mind'. It quoted me as saying that three weeks earlier I had said that I had no proposal to resign my position and certainly not because of the appointment of a Chief Executive. The previous Thursday my position had been confirmed and the *Times* (16 February 1990) had run a story that 'Ridley's team holds key to Eurotunnel progress.' "Today", I said, "it has been announced that I am leaving,"

There had been a row about the role of Morton. It was settled at the highest level — at least temporarily. He had told me some time previously that if one of us had to go, it would not be him. That seemed reasonable, as he appointed me after urging me to join the project on more than one occasion. The press had described me as being between a hammer and an anvil — TML and the chairman. It was said that my diplomacy had prevented a final clash between ET and TML for many months.

"None of this is what the project is about", I said. The real issues of the project were progress in tunnelling and transportation, and the cost of the project. Happily, tunnelling was now going very well, on the British side, as well as the French side. Good progress was being made with the transportation system, though problems remained to be solved. The final cost of the project was much clearer than one year previously. That was what the project was about.

There was bound to be argument. Enormous sums of money were at stake, but we were building one of the most significant transportation links in the world. "It will be built. It must be built. It is the missing link in one of the highest density and most prosperous corners of the world, from Glasgow to Milan".

One question of moment I did not raise. A great deal of Board time focussed on revenue estimates. Every time new money was to be raised the Board had to sign up to the passenger, and therefore the revenue, forecasts. At some stage in a one-on-one discussion with Alistair I did 'wonder' whether I would be able to sign. It did later occur to me that it might be this, rather than a series of other issues that most concerned him.

Clients and contractors either succeed together or fail together — is one of the most important lessons I learned in Hong Kong, but I utterly failed to have my ET colleagues, particularly the Chairmen, understand.

*The Sunday Times* wrote, under the heading, 'Light at the end of the tunnel', on the results of a meeting of the builders, bankers and shareholders in the Channel Tunnel project 'locked in the climax of a tense 10-day poker game' with the stakes the highest possible. 'Would the £7.2 billion project be finished or was it destined to end like the other two (earlier) abandoned tunnels?'

"The contractors wanted someone sympathetic to their problems, and believed that had happened when (he) was appointed project managing director, but it did not work. Ridley was seen by Morton as being too sympathetic to the contractors, while the contractors wanted to calm Morton down. Morton also began to suspect that Ridley could emerge as a rival. On 12 February, ET produced a management reorganisation that made Andre Benard, formerly co-Chairman with Morton, sole chairman of the ET Board. Alistair Fleming, of BP, and Keith Bernard, who was running San Francisco's rapid transit system, came in to the new posts of managing director-construction and managing director-transportation. Ridley remained project managing director." (*Sunday Times*, 25 February 1990)

Eurotunnel's press release (15 February 1990) announcing these changes had stated that Fleming and Bernard would 'give Dr Ridley the support he needs to manage the contract with TML as tightly as is necessary'. But the *Sunday Times* went on,

"There was only one problem: the reorganisation made Morton chief executive with as much potential as before for hands-on involvement and biting criticism. The builders were even more enraged... By 16 February ET and the contractors appeared to have dug themselves into a hole that was threatening to swallow the whole project. With disaster looming, the Bank of England, in the person of Robin Leigh-Pemberton, its governor, stepped in ... Leigh-Pemberton was perfectly suited for his role as concilia-tion and arbitration service to the world's biggest construction project. The bank had been midwife to ET, encouraging international banking support for the tunnel project, rallying sceptical financial institutions to back ET's Equity II and III share issues. Most important of all, in the circumstances, it had been instrumental in recruiting Morton. Leigh-Pemberton's inter-vention broke the deadlock. ET would change the new management struc-ture, only four days old."

Only four days after I had been appointed to an enhanced role, by mutual agreement between myself and Morton, I was out. Why did not I fight? To some extent I was worn out. I was losing much more sleep over ET than I ever did during the 12-month trauma post King's Cross. Although the year after the King's Cross fire was a very demanding time for me and for the organisation I led, at least I had the advantage of knowing very clearly what we had to do. With ET I had a clear idea of what was wrong with the genesis of the project and the ruinous dysfunctional relationships involved on all sides, but I had found no way of improving that situation.

Not only could the chairmen not be persuaded of a different way forward, it also seemed that I had joined the project too late to introduce a root and branch change to relationships. To be fair to Morton, not always easy at the time, even he had probably joined too late to do the job properly, though whether he would ever have wanted to do it properly — by my definition — is a moot point.

Lesson — By mutual agreement is a good way to leave a project.

Fortuitously, but comforting for me, in the substantial coverage in the 22 February *NCE* was another piece by Ty Byrd. Under the heading, 'Ridley walks away with respect', he wrote

> "A large and respectful man approached Dr Tony Ridley at the Association of Consulting Engineers annual dinner last week. (I was talking to Ty at the time). 'I was on the Hong Kong mass transit with you in the late 1970s', he said. 'They were very happy days, on a wonderfully well managed project. If there is truth in your leaving the Channel Tunnel, then God help it'".
>
> "Press Ridley very hard on whether Eurotunnel's chronic problems with its top management will ease with his departure and he smiles wryly but refuses to comment. Nor will he respond to gossip that his relationship with Alistair Morton was every bit as difficult as the Channel Tunnel contractors' still is with Morton. Ridley is nothing if not discreet. Friends and admirers speak up on his behalf however. Those that know him professionally regard him as an extremely effective manager of large transport projects, a good team leader who can be tough and nasty when the occasion merits. Puzzlement is expressed at his inability to get to grips with Morton. 'Must be his terms of reference there', one observer remarked. 'Hope for his own sake that Ridley's successor is given more clout.'"

## 9.3 Aftermath

The rowing and bad news continued long after my departure, as indicated by regular newspaper headlines — 'Why ET is still in a hole' (Independent, February 1991); 'Dispute may stop work on Channel Tunnel' (*Independent* on Sunday 13 October 1991); 'Claims may delay Channel Tunnel' (*Times*, October 1991); 'Builders on brink of split with ET' (*Guardian*, October 1991).

On 23 October 1991, the *Evening Standard* carried a piece under the headline 'TML tirade over ET cost', which said, 'The 10 Channel Tunnel contractors today broke silence to demand that ET shoulder the full £1 billion cost overrun on the project. In an astonishing outburst, the TML contractors insisted that ET, whose chief executive, Sir Alistair Morton, has been the focus of controversy throughout the lifetime of the project, has broken the spirit of the original agreement. Other headlines followed'.

'ET and TML near deal' (*Independent*, December 1991); and 'Court fillip for TML in dispute over tunnel costs' (*Independent*, January 1992); and 'Builder's success puts pressure on ET' (*Times*, January 1992); followed by 'Channel Tunnel delayed until autumn next year', (*Financial Times*, Feb 1992); then 'ET ordered to pay more' (*Financial Times*, March 1992).

In 1993, Morton gave the 10th Lord Nelson of Stafford Lecture at the Institution of Electrical Engineers, 'An April Fool's Guide for British Engineers' (Morton, 1993), which was reproduced in part in Platform in the *New Civil Engineer*. It produced furious responses from two people who had been part of the TML project team. (*NCE*, 22 April 1993). The first, who had HKMTRC experience, said that he had read the Platform article with great interest and not a little incredulity. 'In the area where I principally worked', he said, 'it became necessary to disengage certain work packages, simply to get some procurement moving, while ET interminably debated minor points or different personnel wanted different solutions, all apparently giving no credence to the experience or professional integrity of TML's project teams'.

He quoted Morton, 'TML's planning was so awful that they lost about five months against programme within the first nine months of E&M installation', and commented that if correct, he should be pleased that it was no more, given the difficulties put in its way. 'I could give numerous examples, but will limit myself to the tunnel track-work system, which was in fact taken on board by TML at ET's instigation, but which was still subject to innumerable requests for both concept and design changes (in addition to the perfectly proper material, manufacture and performance testing) over a period of over 12 months'. In the same NCE an editorial took Alistair to task under the heading 'Partial Truth'. In the same edition was 'Chunnel relations sour as accusations fly'.

The *Sunday Times*, on 25 April 1993 had, 'ET heads deep into its worst crisis'. In May Neville Simms, Chief Executive of Tarmac, took over from Joe

Dwyer of Wimpey as coordinating Chairman of the five British contractors and thus of dealing with Alistair. He warned, in the *Evening Standard* (7 May 1993), that the contractors were determined to get their money. 'We are determined to deliver the project and we are determined to be paid'. Significantly he also said, 'To commission and operate the most complicated railway system in the world will require cooperation between the two parties'. At a stroke he had identified, finally, a basis on which ET and TML could form an alliance.

Each of his predecessors — Frank Gibb, Peter Costain and Joe Dwyer — were fine men, but seemed unable to cope with Morton's ways. Perhaps Simms had a greater sophistication, as exemplified by his later directorship of the Bank of England 1995–2002. In July 1993 the *Times* announced, 'ET and TML reach agreement'. In September *Construction News* ran a piece relating that Lemley would leave the project soon after the tunnel was handed over to ET on 10 December 1993 for final commissioning before it opened for business in spring of 1994. He was to be replaced by Haro Bedelian of Balfour Beatty. When asked to comment on Lemley's departure Morton, gracious to the end (!), would only say, "We wish Mr Bedelian all the best".

After I left I was involved in three celebratory events for the Chunnel. On 1 December 1990, TML had organised their own party at Dover Castle to celebrate the meeting of the British and French Marine Service Tunnels under the Channel, to which Lemley invited me. Then early in 1993 I chaired, as VP, a crowded ICE meeting at Methodist Central Hall. (*NCE*, 11 February 1993) In opening, I noted, with some sadness, that the Chunnel should be described as a major transportation project, and an Anglo–French one at that. However we were there to recognise the building of three tunnels between Britain and Continental Europe and honour British civil engineers.

Finally, on 26 February1994, I attended Eurotunnel's historic lunch under the Channel. I was reminded of an earlier, simpler lunch, to celebrate the opening of the Piccadilly line extension to Heathrow.

In the event the Tunnel opened on 6 May 1994, though not in full service, and the financial problems went on. On 20 May 1994, the *Evening Standard* announced, 'ET was teetering on the brink of financial catastrophe today as it emerged that its forthcoming rights issue will need to raise up to £900 million to stave off the latest Channel Tunnel funding crisis'.

**PHOTO 9.2   Celebration Lunch Under the Channel.**

When, later, I became a Professor at Imperial College I taught a number of courses. One was the Principles of Project Management. It was helpful to present a number of case studies based on my personal experience. The star example of a 'good' project was the Mass Transit Railway in Hong Kong. Sadly, the prime example of a badly conceived and badly managed project was ET. These, too, also served to contradict the widely held view that private sector projects are always better handled than those in the public sector. The MTRC was a public body and, at Margaret Thatcher's insistence, ET was in the private sector.

Alistair Morton died on 1ˢᵗ September 2004. The NCE (September 1994) wrote about his manful contribution to the 'project of the century'.

It added 'Morton later took his tenacity and commitment with him to the Treasury's private finance task force where he helped the Conservatives and then New Labour kick start the Private Finance Initiative. His spell as chairman of the fledgling Strategic Rail Authority was less distinguished'

We saw little of each other while I was a full time academic but, both being Past Presidents of CILT, we met at an annual lunch. He told me that he thought he had been a failure in the SRA job. I was amazed. The Morton I remembered would never have admitted to failure, even had it been true. He was one of the brightest, hard driving, tactically brilliant people I have ever been involved with. But strategy was not his strong point, and certainly not inter-personal relations.

## 9.4 Postscript

Nearly 30 years after I joined the project, Eurotunnel is now providing an excellent service, with growing traffic and improving financial performance. Much of the Civil Engineering for the project was first class, especially the tunnelling. More generally UK's score rate of success with complex projects is improving, many of which are now being managed by engineers who were only undergraduates when the Chunnel was being built. So I can't call it a total failure, but it could have been managed very much better, starting 'up front' — not just before I, but also before Morton and Benard became associated with it.

The building of a railway between Britain and France was always going to be difficult, if only because of its complexity, especially the financing. If ever there was a project that epitomised 'there is more to engineering than engineering', this was it.

## Bibliography

ET press release (15 February 1990).
Fetherston D (1997), The Chunnel, Random House.
Henderson N (1987), Channels and tunnels, Weidenfeld and Nicholson, London.
Morton A (1993), An April fool's guide for British engineers. The 10th Lord Nelson of Stafford's Lecture, IEE.
*NCE* (6 April 1989).

*NCE* (1989), Team spirit lifts chunnel outlook, 18 May.
*NCE* (1 February 1990).
*NCE* (22 February 1990).
*NCE* (11 February 1993).
*NCE* (8/15 April 1993).
*NCE* (22 April 1993).
*NCE* (September 1994).
Nichols Associates (1989a), Brief review of the Channel Tunnel project.
Nichols Associates (1989b), Channel Tunnel construction project, Recommendations.
Ridley T (5 April 1989), Memo to staff.

## Press

*Construction News* (18 January 1990).
*Contract Journal* (25 January 1990).
*Evening Standard* (23 October 1991).
*Evening Standard* (7 May 1993).
*Evening Standard* (20 May 1994).
*Financial Times* (13 January 1989).
*Financial Times* (February 1992).
*Financial Times* (March 1992).
*Guardian* (22 October 1991).
*Independent* (26 January 1990).
*Independent* (22 February 1990).
*Independent* (23 January 1992).
*Independent* (February 1991).
*Independent* (December 1991).
*Independent* (13 October 1991).
*Sunday Times* (28 January 1990).
*Sunday Times* (25 February 1990).
*Sunday Times* (25 April 1993).
*Telegraph* (25 January 1990).
*Times* (16 February 1990).
*Times* (14 October 1991).
*Times* (January 1992).
*Times* (28 July 1993).

# Part III
# Changed Perspective 1991 to Date

# 10

# Transport

Chapters 10–13 are drawn from my time in academia and retirement, when there was more time for reflection, during which I was also invited to deliver a number of lectures, on topics of my choice. The material presented is therefore more contemplative than in the earlier case study chapters. Thus, the format of these chapters is different.

In turn, I address Transport (and Imperial College), the Engineering Profession (and ICE and other professional bodies), Project Management and Risk, and International activities. I also include selected passages from lectures, which remain relevant today.

## 10.1 Transport at Imperial

When I took my PhD at Berkeley I fully intended to follow an academic career, but Peter Stott and the GLC got in the way. Being something of a 'hoarder' I have kept detailed accounts of my papers and publications. As a competent public speaker this often led to invitations to address conferences, almost exclusively in the transport field. Many of these addresses were turned into papers. There were, in addition, invitations to write papers for publication. These were hardly 'academic', not being accounts of research, but they did require keeping up with developments in the field and kept my name in front of academic members of the profession. Nevertheless, it was something of a surprise to find myself the holder of the Rees Jeffreys Chair of Transport at Imperial College, for my last full-time position. To become an academic at the age of 57 initially seemed unwise, but having thought about it, I decided to accept the role and have never regretted the decision. My appointment was to run from 1 January 1991.

Imperial College is one of the great technological universities of the world. It was an enormous privilege to have been asked to go there. During the next nine years it became clear that it was also a marvellous way to round-off a career. For the first seven years I focussed on the Transport Section, albeit with the ICE presidency thrown in. It was only when I became Head of Department of Civil and Environmental Engineering that I became fully aware of the weight of talent in other sections within Civil Engineering, and elsewhere in engineering, science and medicine.

Ty Byrd ran a piece in *NCE* (1990), 'Train to Academia'.

> "There is an old saw which runs, 'those who can — do, those who can't — teach'. It is occasionally made use of to annoy academics, and is currently on the mind of Tony Ridley, ex-London Underground, ex-Eurotunnel, and former substantial builder of Hong Kong's MTR."
>
> "The Chair seems to have been built around him, so ideally suited is he for it. Imperial already has a Transport Group within its Civil Engineering Department, one with a long but fading reputation for innovative research into Transport Engineering. Ridley is charged with strengthening the team and — not to put too fine a point on it — re-establishing Imperial as a centre of excellence in Transport Engineering. In his own words, he has to bring together education and research in all the disciplines that contribute to providing solutions to the engineering problems of highways, railways, and airports. He is clearly as pleased as Punch with his new career, long having held strong views on the education of young engineers. 'They have to be taught to achieve things', he says. Achievement is what it is all about."

Thus, although I was joining an academic institution, I wanted to imbue staff and students with the concept of success and delivery, to help Transport Engineering at Imperial, which had nearly collapsed, to develop a reputation for excellence, both at home and internationally.

Transport goes back a long way at Imperial, starting with a post-graduate course in Railway Engineering in 1908 (Brown, 1985). In the early 1960s, Colin (later Sir Colin) Buchanan, a College graduate, was completing Traffic in Towns. (Buchanan *et al.*, 1963) He was supported by a small working group and, when he was appointed to the Chair of Transport in 1963, he brought with him several of the members of that team. A new inter-collegiate course in Transport was initiated in 1981, jointly with staff of the Transport Studies Group at University College London. This collaboration with UCL was enhanced with the creation of the University of London Centre for Transport Studies (ULCTS).

**University of London
Centre for Transport Studies**

IMPERIAL COLLEGE OF SCIENCE,
TECHNOLOGY AND MEDICINE

UNIVERSITY COLLEGE LONDON

**PHOTO 10.1.  University of London Centre for Transport Studies.**

When Buchanan retired in 1972 the Imperial programme was kept going by Geoffrey Crowe until he was due to retire, when the future of Transport at Imperial looked distinctly unpromising. Thanks to the HoD Patrick Dowling, a structural steel specialist, money was obtained from the Rees Jeffreys Road Fund to support a Chair.

## 10.2  The Rees Jeffreys Road Fund and Imperial

The generous support of the Chair for five years was consistent with the Fund's objectives. The Trustees had the wisdom to support a Chair of Transport Engineering, rather than Highway Engineering.

William Rees Jeffreys (1872–1954) the Founder and sole benefactor of the Road Fund that bears his name, opened his historic and autobiographical record of 60 years of road improvement, 'The King's Highway', with the words 'I knew my mission in life'. In 1937, Lloyd George described him as 'the greatest authority on roads in the UK and one of the greatest in the whole world'.

In 1910, he was appointed Secretary of the newly formed Road Board, an office he held for nearly the whole of the life of the Board until its disbandment on the formation of the Ministry of Transport in 1920. A bachelor and also into French Impressionists before they were fashionable, or valuable, he had money to establish the Fund in December 1950.

At the time of my appointment in January 1991, I described the purpose of the Chair and the role of Transport in Civil Engineering in the Imperial College Engineer (Ridley T, 1992a). The success of the Rees Jeffreys venture was later recorded in the report (Ridley T, 1996) to the Trustees. The Fund not only supported the Chair over five years, in 1991 it also made a grant to support the development and re-fitting of the library, which moved from a cramped room with space for only four readers, and for the provision of computer workstations.

## 10.3  Inaugural Lecture ('Transport Matters')

Every professor gives an Inaugural Lecture (Ridley T, 1993a), basically to introduce himself to the university and to explain what the subject is all about. In particular, a new professor will often explain the nature of his research. In my case I could offer thinking and experience, but not research.

Since Transport was not yet fully accepted as a 'real' Civil Engineering subject, I set myself the task of explaining to my new academic colleagues, and indeed to the profession, what the study of transport involved. I chose the title 'Transport Matters', a not entirely novel play on words, and began by saying that transport matters to everyone, from the starving Somali who needs good food distribution to the exporter in the global market place. But what is Transport Engineering? I described it in terms of

- Various characteristics — human, materials, vehicles
- The need for engineers who can command the totality of physical attributes, operations, communication, human resources, finance and funding, organisational and institutional issues and environmental impacts.

"It can be described in terms of various characteristics — capacity, cost, demand (frequently neglected by Civil Engineers who have been sadly supply-oriented), regularity, reliability, safety and supply. Its customers have various characteristics — age, gender, wealth, car ownership and there is a wide variety of types of freight to be moved. A significant part of the skill and training of a Civil Engineer is knowledge of engineering materials. The Transport Engineer has to deal with perhaps the most difficult material of all — the human being. Transport is conventionally, and perhaps unhappily, divided by mode — bus, rail, road, sea, air and may also be described in terms of infrastructure and equipment — aircraft, bridges, automobiles, signalling, tunnels and trains."

"Transport is about the movement of people and goods, and therefore its study, is multi-disciplinary, being carried out in other Sections within the Civil Engineering Department — Geotechnical and Public Health, but also in the Mechanical and Electrical and Electronic Engineering Depts."

"In this country the study of Transport has grown out of Civil Engineering, through Highway Traffic Engineering to Transport Planning and Engineering, gathering up along the way the skills of geographers, economists and others. Transport is about the movement of people and goods and is thus about much more than civil engineering infrastructure. It combines strategies and policies, and processes with infrastructure and equipment. Movement is increasing rapidly worldwide. This increase is leading to concern for the environment."

"The renowned words of Tredgold in the early 19th century became part of the Royal Charter of ICE, 'Civil Engineering is the art of directing the Great Sources of Power in Nature for the use and convenience of man'. This elegant definition may be thought to be somewhat homocentric for today's purposes. It is interesting for those of us in Transport to note that the Charter goes on to say, 'as the means of production and of traffic in states both for external and internal trade, as applied in the construction of roads, bridges, aqueducts, canals, river navigation and docks, for internal intercourse and exchange, and in the construction of ports, harbours, moles, breakwaters and lighthouses, and in the art of navigation by artificial power for the purpose of commerce, and in the construction and adaptation of machinery, and in the drainage of cities and towns'."

"But to set alongside Tredgold's short definition, not to replace it, I prefer a more 20th century definition, 'Civil Engineering is the practice of improving and maintaining the built and natural environment to enhance the quality of life for present and future generations', or less elegantly, Civil

Engineering is about 'getting things done'. If it is, then we must teach undergraduates about economics, net present values, contract law, and the essentials of project management."

"To create successful transport projects, indeed all civil engineering projects, we need engineers who can command the totality of — the physical attributes of a project; operation, communication and human resources; finance and funding; organisational and institutional questions; and environmental impacts. From a Dutchman I had heard this summed up as the 'pentagon' of hardware, software, 'finware', 'orgware' and 'ecoware', to which was added by a Chinese student when I was lecturing to a class of post-graduates, 'you have forgotten luckware!'"

I concluded by claiming that the field of transport, at one and the same time, provides a profound intellectual excitement, and requires a broad range of practical skills in order to 'get things done'. Furthermore it is an ideal subject for Imperial College, established by Royal Charter in 1907 'to give the highest specialised instruction, and to provide the fullest equipment for the most advanced training and research in various branches of science, *especially in its application to industry*'.

As a personal aside, it was many years after I learned of Tredgold's definition that I discovered that he was born in 1788 at Brandon, Co Durham, just a stone's throw from where I was at school for five years.


## 10.4  Carmen Lecture ('Transport and Society: Master or Servant?')

I am a Carman, or rather a Liveryman of the Worshipful Company of Carmen. The Company sponsors a series of lectures at the Royal Society of Arts, and invited me to give the Inaugural Address (Ridley T, 1993b). I set out to demonstrate the challenge and the complexity of transport.

I took the opportunity to tell some family stories that had always fascinated me, as an illustration of how the world has changed over only two generations. My father first travelled from Northeast England at the age of 26 to the 1926 Wembley Exhibition, sailing to the Thames as a passenger on a coal-boat. I first travelled, by car with my parents, at the age of 14. Our youngest son, Michael, had already travelled twice around the world, by air when we lived in Hong Kong, by the time he was 12 years old.

'"Transport, which most of us see as getting ourselves and our goods from A to B, is enormously complex. At any time there is a balance between demand and supply, strongly influenced by price. Demand is influenced by a whole series of determinants — population, employment, land-use patterns, wealth and so on. As demand rises asymptotically towards capacity the flow of movement slows down and, more importantly, reliability deteriorates rapidly.'

'The transport issue for the Channel Tunnel Rail Link was not, as has been represented, a question of saving a few minutes on a journey from Folkestone to London. It was more a matter of how, and at what cost to whom, it was possible to guarantee a reliable journey time for a high-speed train through an already heavily congested existing network.'

'The word 'network' is central. Theoreticians carry out intricate mathematical calculations to describe the flow along a single traffic artery. This is further compounded when the moving vehicles do not have uniform behaviour characteristics and the capacity of the artery is not an absolute number. This is certainly true of roads, but also applies to rail and air transport. In terms of efficiency, and customer satisfaction, there are great advantages in finding a balance in which satisfied demand (flow) at any point in a network is within capacity by a sufficient margin to enable reliable operation.'

'The theory of networks is a branch of applied mathematics in its own right with enormous complexities even when the flow of traffic can be assumed to be uniform. It can be imagined what is involved when the demand characteristics of people and goods are also taken into account, together with somewhat 'elastic' measures of capacity and complex pricing mechanisms.'

'The fundamental questions are not high fares or subsidy, public or private transport, bus or light rail, electric power or diesel, or any of the many other issues that raise such passions among the public, professionals and politicians. In order to address any challenging topic the first question to address is 'what are we trying to achieve?' I am constantly amazed how often people sit around Board and other tables and argue about solutions before they have addressed, let alone answered, the questions.'

'Just like large multi-national corporations, governments must develop and implement strategies — not least to provide the leadership and framework within which private companies, entrepreneurs, private funding and public authorities can get on with the task of 'getting the job done', whatever that job might be.'

'However, it is not enough to think of a strategy alone. I revert to the concept of a 'network', to the systemic nature of all transport, which means that it is most unwise to make decisions in one place, or for one mode, without reflecting on the implications or effects elsewhere. This is pre-eminent in the case of long-term projects, whose gestation periods exceed normal economic and political circles. But it is true also whether we are contemplating bus deregulation, privatisation of railways, pollution by automobiles, road building or whatever.'

'The strategy that emerges from this framework must be clear to all — the public, whether motorist, pedestrian, public transport passenger, young or old, developer, contractor, banker, traffic engineer or retailer. The selling of the strategy is a crucial educational part of the process. At its heart is the need to ensure that everyone understands that trade-offs are necessary. Mobility without regard for environmental protection is no more satisfactory than ecological nimbyism without regard to the crucial contribution of transport to the quality of life.'

'The transport strategy must be rooted in the social and economic policies agreed for the area. A transport strategy that is in conflict with an area's social and economic policies is not likely to be a great success. The outcome is likely to be a preferred combination of public transport and road investment, the use of electronic means of guiding or controlling the user of parking and road space, mechanisms to restrain the use of road vehicles, and means of managing both the provision of public transport and whatever means of restraint are to be used.'"

Looking back, as I write now, I feel sure that these words about strategy were strongly influenced by experiences, good and bad, over the years — the lack of any realistic strategy at the time of the Motorway Box, the excellence of the processes we were required to follow in Tyne and Wear, the soundness of the thinking by the Hong Kong government that underlay the Mass Transit Railway, and my frustration that it became necessary to give up my task with the Underground just when we were facing up to the implications of a 60 percent increase in ridership.

## 10.5  Railway Technology Strategy Centre

An important development within the Transport Section arose when we were able to add a railway specialism, as a result of a meeting in 1992 with Peter

Watson, the new Board Member for Engineering at BR. His concern was that the post did not 'command' engineering resources. The civil, mechanical, electrical and signalling engineers did not report to him. He wanted to have a brain-storm in order to define his job. After several discussions we came to the view that central to the task was to define a technology strategy for BR. Out of this came an agreement for BR to support a unit within Imperial's Transport engineering section, the purpose of which was announced in a press release on 14 July 1992 — 'New Railway Technology Strategy Centre at Imperial College'. This said that initial support had been provided by British Rail, for a minimum of two years, to work closely on the assessment of technical priorities, methods of monitoring, and developing and managing technology in comparison with other railways and industries. The aim was to provide input into BR's technology strategy for the future. The new Centre would form part of the new ULCTS that had recently been established by Imperial and University College to develop inter-collegiate research and teaching.

Patrick Dowling expressed gratitude to BR for adding to Rees Jeffreys' contribution and said, 'we are now beginning to see our hoped for development covering all aspects of transport'. This was an important point. On my arrival the Transport section's reputation was based on excellence in everything to do with roads and traffic, but little else. Ten years later, as well as Rail Transport, Air Transport and Ports had been added to its programme. With the establishment of RTSC, to add to other on-going activities, it was safe to say the survival of the Transport section, and my appointment, had now been justified.

The first piece of work for Watson was to help him to decide the future of the BR Research Centre in Derby. This was followed by the task that we had discussed when we first met — to develop a framework for BR's Technology Strategy (RTSC, 1992), this was followed by a review of the technical options and business case for the proposed West Coast modernisation (RTSC, 1993), and a technology audit for BR (RTSC, 1994).

Meanwhile several of us who had links with both LU and HKMTRC began to discuss how we might develop studies bringing together knowledge from both of these organisations. Building on my work with UITP, on productivity comparisons between metros, described in the Tube Chapter 6, we put together a programme for a series of benchmarking studies for which metros would be the clients, but for which RTSC would gather and analyse relevant data. We obviously wanted more than two metro members. London had a long history of working with RATP in Paris, so they were included, which meant that Berlin should also be included. We did not want just Hong Kong and Europe, so New York was added. So the whole programme mushroomed. Soon there were four benchmarking groups — CoMET and Nova for metros, one for suburban rail, and one for bus benchmarking.

The success of RTSC and its benchmarking work is a joy. The geographic spread of membership is mind-boggling. The members, at my latest count, are CoMET (Beijing, Berlin, Guangzhou, Hong Kong, London, Mexico City, Madrid, Moscow, New York, Paris, Santiago, Sao Paulo, Shanghai, Singapore and Taipei), NOVA (Bangkok, Barcelona, Buenos Aires, Brussels, Delhi, Istanbul, Kuala Lumpur, Lisbon, London DLR, Milan, Montreal, Nanjing, Naples, Newcastle, Rio de Janeiro, Sydney and Toronto), and Buses (Barcelona, Brussels, Dublin, Lisbon, London, Milan, Montreal, New York, Paris, Singapore and Sydney).

Because of the broadening of the role of RTSC over the years we eventually changed its name, while keeping the initials that had become known around the world, to Railway and Transport Strategy Centre. The Transport section has gone from strength to strength ever since. So much so that, by early in the 21$^{st}$ century, there were already four Professors of Transport within the Department of Civil and Environmental Engineering. The BR support was to last for three years after which we had to ensure that it had become self-supporting. It has done and was still flourishing 20 years after the meeting with Peter Watson in 1992.

## 10.6 Sir Dugald Clerk Lecture ('Light Rail: Technology or Way of Life')

At about this time I had been invited to give ICE's annual Sir Dugald Clerk lecture, which was instituted in 1938 following a bequest made to the Institution by Clerk who was a Vice-President. The lecture seemed particularly well timed as it was the first public lecture (Ridley T, 1992b) following the appointment as Professor and I had just completed nearly 20 years as the President of the Light Rail Transit Association (LRTA).

I decided to use the lecture to make the case for Light Rail (modern trams), which were highly regarded on the Continent, but rarely considered in the UK at that time. It had a long history of technological development and success but, in addition if seen in totality, it would help to transform the urban way of life. I emphasised that, in the UK, facing an urban transport crisis, the way ahead was neither to recreate vast bureaucratic authorities nor to hope that the market would miraculously solve the problems. There had to be a vision and a strategy which would include Light Rail, but Light Rail was one tool among the packages of measures available to tackle urban transport problems. The need to encourage transport enthusiasts to look beyond their technology, to address land-use issues as well, to look at 'the totality' of their projects was also underlined.

The lecture was a considerable success, both in London and when I repeated the presentation 10 times, starting in Glasgow.

## 10.7 Head of Department

More than six years into my tenure Ron Oxburgh, the then Rector, invited me to become Head of Department. Colleagues tended to avoid 'administrative' jobs which 'interfered' with their research and, additionally, managing academics was like 'herding cats in a sack'. I didn't have a research profile to protect and, in spite of protesting my age, I was persuaded to do the job until my retirement date in 1999.

## 10.8 Professor Chin Lecture ('What Is a Successful Transit Project?')

After I had become President of ICE an early invitation to travel overseas came from Malaysia. Professor Chin Fung Kee was one of their most eminent engineers. After his death in August 1990 his colleagues created the Annual Professor Chin Memorial Lectures series, to be given in September/October each year and published in December. In 1996 they included each of the first five lectures, 1991–1995, in one volume. Edmund Hambly had accepted an invitation to give the 5th lecture, during his presidency, but died before he could do so. Consequently, I was invited to step into his shoes. I did not give his lecture on his behalf, but wrote my own. Consequently

Edmund's lecture (Hambly, 1995) and my own (Ridley T, 1995a) appear together in the document.

I decided to address the question of 'success'. Too many engineers define success as it relates to hardware, to 'things'. I sought to make the customer, or customers, central to our objectives.

I asked — who is the customer?

"If I am running a public transport system which is heavily subsidised by government, then I am well advised to recognise that the government is a customer to whom I should pay due regard, just as I do to the customers who are also known as passengers. I can often upset both. My passenger customers will not like it when I increase fares. Equally my government customer will not be happy to be told that the system, which he has entrusted to me, desperately needs more investment if it is to perform properly."

I spoke about the 'passenger customer' not least because, in the past, railways had been dominated by engineering thinking.

"Public, or collective, transport will only compete with individual transport if customers are seen as individuals who happen to travel together in large numbers. In my academic life, and with my civil engineering background, I look at the materials used by engineers. Steel, concrete, wood and ground are increasingly being joined by new composite materials. Once again I said, 'but transport engineers must understand the most complex 'material' of all — people' as earlier chapters have said."

The emphasis had moved from production to the market and now to service, which is essentially a process and, in contrast to a product, service cannot be stored since it is produced and consumed at the same time. Another characteristic of a service is that the customer becomes part of the service process. It has taken some considerable time for engineers, indeed many operators, to comprehend the crucial differences between production and service. With concentration on production we tend to ignore demand questions in the economic equation. Efficiency of supply is important, but an understanding of the delicate balance between supply and demand is crucial. The production mentality concentrates exclusively on production measures — car kilometres, cost reduction, staff numbers.

'"The task of engineers, or for operators for that matter, is to get things done. Of course that is not enough. We must get things done right (or well) as well as get the right things done. These we call, respectively, efficiency and effectiveness. In my country politicians of one colour have tended to concentrate on the former and politicians of another colour on the latter. The professional however must perform a balancing act between them which means that we must first think clearly and have a clear understanding about the question — what are we trying to achieve?'

'There is a tendency to confuse efficiency and effectiveness. Service efficiency is often the primary concern of our government customer, while service effectiveness is the concern of our passenger customer. But our passenger customer actually wants both, which I call cost effectiveness. There is no merit in a good effective public transport system which costs too much, nor in efficient public transport which is not effective.'

'If we are to focus on the customer we must be efficient and effective and provide both the government and the passenger a service which is cost effective in that it provides a good service which meets customer needs at an affordable price — in terms of fare or subsidy, or both. It will thus give value for money or, to use an engineering term, it will be 'fit for purpose'. Thus a cost effective service has attributes of both service efficiency, the level of service output relative to the resource cost of supply of inputs, and service effectiveness, the demand for use of the service relative to the level of service output.'

'There seem to me to be five prerequisites for the development of an urban rail system — political consensus, a well-conceived strategy, a small planning team, a single minded implementation team, and easy access to right-of-way. I know of no successful rail development where there was not political consensus. This is much more than the need for political support. Because the creation of an urban rail system is an exercise in the management of complexity, it is well-nigh impossible to carry this out if there is not general agreement across the political spectrum.'

'Every project has problems. That is virtually the only certainty about projects. When the problems arise it is terribly easy to be blown off-course. Take for example Eurotunnel's regular need to return to its bankers, and sometimes its shareholders, when pre-agreed limiting financial ratios relating anticipated revenues and capital costs were transgressed. The discussions with the bankers, more than 200 of them, were clearly necessary but did have considerable impact on the management of the project. In Tyne and Wear the legislation setting up the (political) PTA and the (professional) PTE required them to produce a joint policy statement within 12 months of their creation. This strategy, and particularly the intention to create an integrated urban public transport system, stood the PTA and PTE, and the Metro, in very good stead when the tough times came along later.'

'The planning team must be small though, of course, including a breadth of skills to address all aspects of the project. Large planning teams are such a management challenge in themselves, particularly when newly created, that it is very easy — indeed almost inevitable — to lose focus and thus produce an unsatisfactory strategy or plan. Once a project is identified, based on a sound strategy, well planned and attracting a political consensus, it must be pursued by a ruthlessly single-minded implementation team. Time is money, as we know. But it is not just a question of being tough. There are a number of well-established disciplines which, if followed, will considerably reduce the chances of the project going awry. Surprisingly the professional project manager has great difficulty in persuading others of the wisdom of these disciplines. The principles of project management are unique in being, at one and the same time, patently self-evident, yet frequently and disastrously almost wholly ignored. At least this has been the case in the past.'

'One of the simplest basic principles is that there is a triangle of time, cost and performance for every project. Of course it is to be preferred if equal weight can be given to all three. But, not only is that not always possible, there is rarely unanimity among all the parties involved in the project — directly, indirectly and externally — about what the balance between the three should be. The project life cycle should be clearly understood. At the simplest level the stages of a project are — project preparation, conceptual (feasibility), development and detailed design, procurement, construction, commissioning, and operation and maintenance. So too should various key roles be understood and separated. The client is accountable for investment in the project and the achievement of the aims of the project. It is the task of the project manager to plan, organise, staff, control and lead the project from start to finish, however defined.'

'To put a new project in place requires knowledge and command of operational, communications and personnel matters; physical equipment and infrastructure; environmental effects; finance and funding; organisational and institutional matters. Not only is each one of these important in its own right. It is the inter-relationship between them which raises the greatest problems. Nearly all engineering problems in the design and development of any system arise at the interface. At a larger scale it is at the interface between the five elements of the pentagon that the difficulties arise.'"

Having signed me up for the Chin lecture, IEM also persuaded me to give a further lecture while I was in Kuala Lumpur, 'Civil Engineering into the Next Century', on the next day (Ridley T, 1995b).

## 10.9 Singapore

It is the practice for academics to serve as External Examiner for courses at other universities. It is encouraged, if only to ensure that external examiners can be found for their own courses. I fulfilled the role four times.

In January 1998, and not long before the end of my academic career, I was invited by Henry Fan, the Director of the Centre for Transport Studies at Nanyang Technological University, to take up the role of External Examiner for the Master of Science degree in Transportation Engineering. This involved a visit to NTU once during the two-year term of appointment to assess the course curriculum. I said yes. The two-year appointment was extended for a further two year until 2002.

Meanwhile, in January 2000, Fan had written with the news that Nanyang had received a donation from Singapore MRT Ltd. for the establishment of the SMRT Professorship in Transportation Studies. He asked if I would be the inaugural holder of the position. 'It would probably involve a stay of two to four months at Nanyang in the second half of this year', he said. It did interest me but, although now an Emeritus Professor at ICL, I was rather busy and a few months seemed too long.

It was possible to commit to six weeks, but that was agreed with alacrity. Life was complicated because I had accepted an invitation as the keynote speaker at a symposium in Shanghai (TMR, 2001a) during this time, and also a separate CILT meeting in Singapore, when I stood down from the Presidency.

The invitation to speak in Shanghai came from Joseph Chow, the President of HKIE, and Co-Chairman of the Shanghai Symposium. At the same time I was pressed to give the Faculty of Engineering 90th Anniversary

Distinguished Lecture at Hong Kong University (Ridley T, 2001b). I gave three public lectures during my time at Nanyang, on Cars and the Environment, Benchmarking, and Risk (Ridley T, 2001c). The first was published in the *Journal of the Institution of Engineers*, Singapore (Ridley T, 2002) and the second, in extended form, by the LTA (TMR, 2008), which can be found in Chapter 11.

My last formal contact with Singapore came when I was invited, in 2007, by the Minister of Transport to join a new International Advisory Panel and served for three years. Finally, in February 2011, I was pleased to accept an Honorary Fellowship of the LTA Academy.

## 10.10  RAC Foundation

In 1992, I had been invited to join an Academic Advisory Group of the RAC Foundation for Motoring and the Environment. The members of the Group, under the chairmanship of Professor Sir David Williams, a lawyer and Vice-Chancellor of Cambridge, included Malcolm Grant, Professor of Land Economy at Cambridge and subsequently Provost of University College London; Brian Hoskins, (then) Professor of Meteorology at Reading, and now Director of the Grantham Institute for Climate Change at Imperial; and Tom Williams, Professor of Civil Engineering at Southampton — my former concrete lecturer; and myself. Given that my career had focussed heavily on urban rail, this gave me welcome insight into the issues relevant to Highway Engineering and strategy.

## 10.11  Cars and the Environment: A View to the Year 2020

The executive summary of the Group's first report (RACF, 1992) said,

> "'In the light of growing concern about the relationship between the steady increase in levels of car use and environmental damage, the RAC decided in 1991 to establish an independent charitable foundation to address the problem. The aim of the RAC Foundation is primarily to seek out, define and promote, by the encouragement and sponsoring of specific research projects and initiatives, those policies and programmes which will reconcile the need and desire for personal powered mobility with the protection of the environment.'

'In order to ensure that its work will be securely based, the RAC Foundation resolved to bring together a small group of academics expert in fields relating to motoring and the environment, The Academic Group was asked to review the level of knowledge and understanding about cars and the environment, assess the relative importance of different aspect of the problem, and make recommendations for future research.'"

This was the RAC before the Club sold its Motoring Services arm. It was also the RAC that had the reputation of having a number of 'mad motorists' among its membership. I regard the document that was produced as being an excellent piece of work, which not only set the parameters for discussion of the issues but also established the RAC Foundation as being a responsible body that carries out good quality studies (Brendon, 1997).

Our conclusions noted that our terms of reference had set the clear objectives of reconciling the need and desire for personal powered mobility with protection for the environment. Achieving this objective was important both because of the growth of concern about the environment, and because of the extent to which our society had become genuinely dependent on the car in particular. It was also so obviously the preferred choice — and indeed an aspiration — of so many people to acquire and retain the enhanced personal freedom which the car provides, There was an obligation in a free society, to try to allow an aspiration such as this to be met to the greatest extent possible without bringing unacceptable costs upon society as a whole or on other individuals.

We considered that one of the most important conclusions to have emerged from our work was the need for greater recognition and understanding of the extent of dependence on the car, and the valuation of benefits, as well as the dis-benefits that the car had brought. Without a systematic and balanced approach to this, the basis for decisions about the future development of personal powered mobility, and the extent and nature of its use, would be flawed.

Much of the report consisted of a carefully considered catalogue of the different environmental effects. A further significant part of the report was devoted to technological change, the assumed need for which implied that the problems associated with cars, in their current numbers and form, were severe.

The most telling comment in the report was, 'Perhaps most difficult of all is the judgement of how severe the problems are likely to become. ... There are a number of factors that have helped us to come to the judgement that the moment has come for the precautionary principal to be applied. While we have concluded that there are very good prospects for the emergence, in the relatively near future, of cars that are remarkably clean in terms of local level pollutants, the $CO_2$ problem still looks to be unresolved'.

## 10.12  Motoring towards 2050

When, in 1999, Motoring Services was sold by its owners, the individual members of the Club, they granted the Foundation a legacy to guarantee its long-term future. The Foundation became a registered charity independent of both RAC Motoring Services and the Royal Automobile Club, at which point I became a Trustee, having already joined the Public Policy Committee in 1997.

In 2000, government was pursuing a policy of raising petrol taxes faster than the rate of inflation when the 'fuel protests' frightened them into retraction. This led Sir Chris Foster, the Foundation's chairman, to write to Prime Minister Blair to suggest that government should conduct a long-term study looking at the future of motoring within transport policy over the subsequent 50 years. Blair replied and said, in effect, what a good idea — why don't you do it? Hence 'Motoring towards 2050' (RACF, 2002), an 18 month inquiry based on input from an experienced steering group including Sir Nick Monck (formerly Permanent Secretary of Employment and second Permanent Secretary of HM Treasury), Richard Ide (formerly CEO of Volkswagen UK), the Secretaries General of the National Society for Clean Air and Environmental Protection, and of the Royal Town Planning Institute, myself and others.

Blair provided the Foreword, in which he said,

"This report cannot and does not represent government policy. But it is a well-argued and interesting contribution to the debate, and particularly so as it is from an independent inquiry by a respected motoring organisation. Meeting the needs of modern motorists, whilst fulfilling our responsibility to protect the environment, is one of the biggest challenges faced by any government. The RAC Foundation first suggested an independent inquiry into motoring issues in October 2000. In replying, I said that we would welcome report bringing greater clarity and understanding of the issues and choices. Eighteen months on this Report more than meets that challenge."

I greatly enjoyed the experience of involvement in a very worthwhile exercise which confirmed my view that, in these days when retirees are mostly living longer and healthily, one can continue to make a contribution before limbs fail and the mind (possibly) becomes addled. Of course, it is a matter of judgement at what point one begins to overstay one's welcome. Retirement from 'retirement' must be well judged.

The Foundation has a long list of other excellent publications, including 'Roads and Reality' (RACF, 2007) in which I was closely involved, 'The Car in British Society' (RACF, 2009), and 'On the Move' (RACF, 2012).

## 10.13 Paviors Lecture ('Transport Engineering: A Future Paved with Gold')

While I had been HoD at Imperial I was responsible for making it possible for the Worshipful Company of Paviors to return from City University to their original home for the annual Paviors Lecture. I received a copy of 'Children of Stones — a history of the Company', with a signed acknowledgement from the author and then Master, Ian Dussek, 'Tony, with best wishes and thanks for initiating the 25th Paviors Lecture at its rightful home' in the year 2000.

I was subsequently invited to give the 2005 Paviors Lecture in the Skempton Building, which was published in Civil Engineering (Ridley T, 2005). The Lecture is specifically directed at both colleagues and students, and I emphasised that I had chosen my title, not in order to define a 'rich life' as a noble aspiration, but because it is fulfilling.

## 10.14 London School of Economics

My last substantive piece of university teaching arose shortly before I came to the end of my full-time involvement at Imperial, which came about because of my involvement with Arup and Jack Zunz in particular. Zunz was one of the great engineers of our generation, but my contacts with Arup went back long before I met him. They were designers of the Byker Bridge for the Metro and had a number of appointments for the MTR in Hong Kong, when my principal contact was Povl Ahm.

Zunz, in retirement, became Chairman of the Ove Arup Foundation. Shortly before I became HoD at Imperial I learned that he had offered the

Department financial support to allow the appointment of a part-time Chair in Civil Engineering Design. They had turned it down and I was appalled. I told him, in confidence, that it was likely that I would shortly become Head, and urged him not to take his money away. In the event he did finance, the role, and we appointed Chris Wise, an Arup engineer, who subsequently went on to cofound Expedition Engineering. Most of the students found him totally inspirational, an opinion not matched by all of his university lecturer colleagues, for many of whom he had a low regard because of their lack of design and construction knowledge, albeit they might be very clever researchers.

The motivation behind the funding was to boost the emphasis on 'the art of engineering', to complement the science that underpins it. Out of this grew the Constructionarium (Ferguson, 2016), whereby students are taught, and can gain experience of, the interface between theory, design and practice. More recently the Constructionarium has moved to the National Construction College in Norfolk, where undergraduates experience what has been called 'a massive field trip with a difference', working in teams with individual students holding various roles, from Project Manager to Health and Safety Officer and construct mini-versions of the Gherkin, the Millau Viaduct, a nuclear power station and many other structures. My role in all of this was strictly limited because of my retirement, but it forcefully reminded me once again what one can help to achieve by making the right facilitating decision in a timely fashion.

At the same time Zunz was developing the ambition to support a programme of urban design, which he succeeded in doing at London School of Economics. The Cities Programme 'connects urban design to urban society. The Programme, located in the Department of Social Policy, is an international centre open to students, architects, engineers, scientists, public sector workers and leaders in the private sector. The programme teaches, conducts research, offers consultancy services, and presents public events.' I began my teaching in 1998, before I retired from Imperial, and continued until 2001. There were three professors — the American Richard Sennett, a professor of sociology; Ricky Burdett, an architect; and myself.

The first year was something of an experiment and what the students made of the mixed messages from the three of us I was never quite clear. Zunz was satisfied with what he had helped to create, but somewhat distressed by the paucity of engineers who had signed up. Also on the staff were Tony Travers, well known in the media as the leader of LSE's Greater London Group, and Kathryn Firth who provided the glue for the programme by running the design studio where the multi-disciplines came together.

When I started at Imperial I encouraged the Masters' students in an interest in engineering history, including writing essays about the Buchanan report and its impact. I was able to repeat this approach at LSE with essays, among others, on the heroic work of Chadwick and Bazalgette in Victorian London.

## Bibliography

Brendon P (1997), *The Motoring Century — The Story of the RAC*, Bloomsbury Publishing, London, UK.

Brown J (1985), *A Hundred Years of Civil Engineering at South Kensington*, Imperial College, London.

Buchanan C *et al.* (1963), *Traffic in Towns*, HMSO, London, UK.

Ferguson H, (2016), *Constructionarium — Making to Learn*, Pureprint, East Sussex, UK.

Hambly E (1995), The risks of daily life, in *Professor Chin Fung Kee Memorial Lectures 1991–95*, Institution of Engineers, Malaysia.

NCE (1990), Train to academia, 8 November.

RAC Foundation (1992), Cars and the environment — a view to the year 2020.

RAC Foundation (2002), Motoring towards 2050.

RAC Foundation (2007), Roads and reality.

RAC Foundation (2009), The car in British society.

RAC Foundation (2012), On the move.

Ridley T (1992a), Transport on the move, *Imperial College Engineer*, Vol. 12, No. 2, pp. 19–20.

Ridley T (1992b), Light rail — technology or way of life (Sir Dugald Clerk Lecture), *Transport Proceedings*, ICE, London, Vol. 95, pp. 87–94.

Ridley T (1993a), Transport matters (abridged), Civil Engineering, ICE, London, Vol. 97, No. 4, pp. 146–147.

Ridley T (1993b), Transport in society: Master or servant? Worshipful Company of Carmen Lecture, *RSA Journal*, Vol. 141, No. 5437, pp. 185–196; *Proc. CIT*, Vol. 2, No. 3. pp. 30–42.

Ridley T (1995a), What is a successful urban transport project? 5th Professor Chin Memorial Lecture, in *Professor Chin Fung Kee Memorial Lectures 1991–1995*, Institution of Engineers Malaysia.

Ridley T (1995b), *Civil Engineering into the Next Century*, Institution of Engineers, Malaysia.

Ridley T (1996), The Rees Jeffreys chair in transport engineering at Imperial College, Report to the Trustees of the Rees Jeffreys Road Fund, 7 June.

Ridley T (2001a), Urban transport – what my experience tells me, Keynote Address, Shanghai-Hong Kong Symposium on Science and Technology — Urban Transportation, Shanghai.

Ridley T (2001b), The globalisation of urban transport, Faculty of Engineering 90th Anniversary Distinguished Lecture, University of Hong Kong, Hong Kong.

Ridley T (2001c), 2001 Public Lectures, SMRT Professorship in Transportation Studies, NTU, Singapore.

Ridley T (2002), Cars and the environment — towards a sustainable future, *Journal of Institution of Engineers,* Singapore, Vol. 42, No. 3, pp. 6–11.

Ridley T (2005), Transport engineering: A future paved with gold? Paviors Lecture 2005 (abridged), Civil Engineering, ICE, London, Vol. 158, No. 2.

Ridley T (2008), Benchmarking — productivity and service, in *Journeys*, Land Transport Authority, Singapore, No. 1, pp. 34–43.

RTSC (1992), A recommended framework for British Rail's technology strategy (unpublished), Centre for Transport Studies (CTS), Imperial College, London.

RTSC (1993), Review of the technical options and business case for the West Coast modernisation project (unpublished), Centre for Transport Studies (CTS), Imperial College, London.

RTSC (1994), A technology audit and strategic input for production services division of British Rail (unpublished), Centre for Transport Studies (CTS), Imperial College, London.

# 11

# The Engineering Profession

## 11.1 Presidential Beginning

This chapter discusses the profession of engineering against a background of experience of membership and the presidency of ICE, supplemented by experiences as President of a number of professional organisations, including the Chartered Institute of Logistics and Transport 1999–2001 and the Association for Project Management 1999–2003. ICE is the most prestigious, if only because it was the first. It was founded in 1818 (its 200th anniversary will be celebrated shortly), and Thomas Telford became its first President in 1820.

Apart from going through the normal process of becoming professionally qualified as AMICE (now MICE), I had devoted little time to my engineering institution. Indeed any professional involvement had focussed much more on the Chartered Institute of Transport (CIT — now CILT).

ICE Past President Peter Stott (1989–1990), one of my mentors, whose earlier impact on my life had been to bring me home from California to work for the new GLC, was concerned that ICE did not give adequate recognition to the transport professionals in membership, nor did it do enough to encourage transport specialists to join. As a result he pressed me to put my name forward as Chairman of the (Institution's) Transport Board. This required me to be elected to Council.

I did as he suggested and was duly elected, starting in November 1990, two months before I took up the Chair of Transport Engineering at Imperial. I became President in April 1995, 4½ years later — possibly the fastest elevation in history? Without Stott's intervention I could not subsequently have become President, a term on Council being a prerequisite.

## 11.2 Whither Civil Engineering?

It had become the tradition in ICE for a senior member, sometimes the Senior Vice-President, to develop a particular aspect of Civil Engineering prior to becoming President. For some years previously the focus had been on Infrastructure. In 1981–1982, a group was established, the Infrastructure Policy Group whose aim was to keep under review the state of the country's infrastructure, and to draw to the attention of decision makers the measures necessary to maintain and improve infrastructure. The topic in 1986–1988 was Urban Regeneration, followed by Congestion (1989), Pollution and its Containment (1990), and Infrastructure — the challenge of '92 (1992). When my turn came, in 1994, my objective was to widen the discussion. I believed that it was time to consider where the Civil Engineering profession was going.

I have already referred to my admiration for Hardy Cross, whose name was well-known to my generation. Bruce Beck, a colleague in the Department at Imperial, is not nearly so well known, but he also had a very important impact on my thinking. He had written, 'Civil Engineering — 2020' (Beck, 1988), in which he described his purpose of initiating a research programme designed to provide the Department with a fundamental appraisal of the long-term future directions of Civil Engineering. He quoted Muspratt (1986) in calling for a high-tech thrust representing the combination of Civil Engineering with computers, and a government and public policy thrust representing the combination of Civil Engineering with management-oriented education. Beck had not succeeded in launching his project and, although he had developed his ideas he was still, in 1992, with the help of Patrick Dowling, trying to get funding. But his life had changed and he was preparing to take a Chair at the University of Georgia, USA, starting the following year.

Influenced by Beck, I wrote a discussion paper, 'W(h)ither Civil Engineering!' (Ridley T, 1994) which suggested that we had to decide where Civil Engineering was going, or it would be in danger of becoming marginalised. There were several possible reasons — if Civil Engineering focussed only on new construction, failed to keep pace with new technologies, could not improve its productivity, was viewed too narrowly, or if it could not adjust to socio-political change.

A Steering Group was assembled with a number of significant, but different, eminent members of the profession to examine the subject. It led to a series of powerful but positive debates about the issues, not least those relating to the role of project management. They included Past President Alan Muir Wood and Martin Barnes of New Engineering Contract fame. The title of the report was the same as for my Imperial paper, with the exclamation mark replaced by a question mark. Although we started in November 1994, with the death of Edmund Hambly it was delayed and the report was not published until 1996 (Ridley T *et al.*, 1996).

The Preface said that (in the mid-1990s) it was increasingly fashionable to look ahead to the new millennium. The inspiration for the report was the thought that it was time for the profession to look ahead, into the next century, to the year 2020, when current twenty-something civil engineers would still not be at the peak of their careers. It had occurred to me, as a professor at Imperial, that my career would not 'end' when I personally reached 65, but about 2040, when the current undergraduates would themselves reach retirement — an important awareness of the influence of educators on their pupils. After all, Fisher Cassie's influence lasted at least until 2000 through me and other of his students.

The report emphasised the need for change as one of its main driving forces. The profession would be influenced by wider political, social and economic trends over which it had no control. Although the report was addressed to UK-based civil engineers, many of them would be working internationally. On a global scale, sustainable development would become the most widespread and far reaching influence on the profession. Construction industry trends would lead civil engineering into a new and untested world. The growth of alternative sources of finance would demand a far more proactive and commercially oriented approach than we had been used to. There would be increasing opportunities to cross the divide between contracting, consulting and municipal engineering as the industry restructured itself.

Political changes also offered an opportunity to reassess and re-invent the role of Civil Engineering in meeting society's needs. Building consensus among all interested parties would become an increasingly important element of this role. To enhance our value to society, we also had to maintain an involvement in all stages of the life cycle of our products and services. To improve the quality of our work, while operating within a commercial context, we would need to address potential conflicts between 'value for money' and short-termism. We would need to show that we could combine short-term profits with long-term benefits. At the same time we would need to develop an ever-closer relationship between sustainability and engineering. This would require an enhanced understanding of ecological principles. It would also provide an opportunity to enhance profitability and value for money.

Sustainability, ethics and acceptability would become closely interlinked themes within our work. We must therefore take the lead in setting ethical standards in our areas of responsibility. Working in an international context would be a major part of the experience of UK-based civil engineers, but opportunities to do so would be constrained by the growth of indigenous skills in emerging economies and lack of finance in struggling economies. We had not to lose sight of the essence of creative and successful civil engineering, which lay in the interaction of design and project management. While the former had not to be reduced to technical analysis, the latter had not to be reduced to administrative control of uncoordinated inputs. Risk management would become an increasingly important aspect of producing optimum solutions. Not least among the reasons for this would be a growing awareness of financial risks, as a result of private sector financial involvement.

An immediate priority was that of transforming the industry from its then conflict-ridden state. Although mechanisms for doing so were available, we would need to focus more explicitly on organisational and human issues in order to make the best use of them. We would need to be more adept at capturing the benefits of innovation, research and technology development to provide added value to our products and services and to remain competitive in a commercial environment. In adapting to change, civil engineers would need to face and resolve several potential conflicts. These included maintaining a healthy balance between competition and cooperation, meeting not only the needs of private sector clients but also wider social objectives, reducing costs while increasing productivity and quality, and extending our technical expertise while enhancing our awareness of the wider context.

The report concluded with an invitation to our successors in 2020 to revisit the conclusions and to look back and determine to what extent our perceptions had proved correct, or not, as part of their own process of looking to their own future. The odds might not be good but, God willing, I might even be around to join in!

## 11.3  Taking over the Reins Early

While still Senior Vice-President, in March 1995, I was in Jakarta on one of my visits as an Advisor to B J Habibie, later President of Indonesia, but at that time Minister of Science and Technology, when I learned that the current President, Edmund Hambly, was very ill and had asked that I should give his address at the Annual Dinner at the Grosvenor Hotel on the next Thursday. I immediately agreed.

By the time I got back to Heathrow, Edmund had already died. Thus, prior to the dinner I had to brief a meeting of all VPs, Directors and table hosts (many of the guests were unaware of the news when they arrived at the Grosvenor), and I was also given authority to act as President until the future could be regularised.

On 1 April, the Council of the ICE met and appointed me as its 131st President, the term to run to 5 November 1996, the original finish date. Thus, the presidency would run for some 19 months. There were two precedents, on the deaths in office of President Gourley in 1956–1957 and President Hartley in 1959–1960. On each occasion the Senior Vice President (Whitaker and Manzoni, respectively) had served out the incomplete term followed by his own year. Interestingly, Sir Ralph Freeman (Past President) pointed out that both occurred during his time on Council, and confirmed that the SVP had taken over in each case (Freeman R to T Ridley, 1996).

## 11.4  Presidential Address ('Is Our Civil Engineering Too Small?')

Every President gives a Presidential Address, when the predecessor hands over to him in November. Because of Edmund Hambly's death, I had already been President for half a year when I gave my Address. Every President is different, and there is a variety of approaches that a new President chooses for his Address. Several of my predecessors had described their careers and/or fields of specialisation. I did not choose to follow them. One year previously, when thanking Stuart Mustow for his Presidency I had remarked that his task was to have carried the baton for the Institution for a year. On reflection it now seemed appropriate to use the analogy of the 'bearer of the torch', rather than 'the carrier of the baton'.

The address was called 'Is Our Civil Engineering Too Small?' (Ridley T, 1995) I reflected the words of Phillips (1961), in Your God Is Too Small, 'the trouble today is that they have not found a God big enough for modern needs. While experience of life has grown in a score of directions, and mental horizons have been expanded to the point of bewilderment by world events and by scientific discoveries, ideas of God have remained largely static'. Thus, I followed my constant theme — about the breadth of civil engineering. Faith and works, research and teaching, the academic and practice, like love and marriage going together like a horse and carriage, you cannot have one without the other. Yet we were all too frightened. We clung to our specialisations, almost like a comfort blanket.

We know that the whole life cycle of a civil engineering project must be addressed if we are to make wise decisions to proceed — planning, finance, design, procurement, construction, commissioning, operations and maintenance, and decommissioning. In five years in the nuclear power industry when I was young I never once heard the word 'decommissioning'. In the past there had been a tendency to concentrate on the design stage. In fact to put a project in place, to get things done, the right things (effectively) done well (efficiently), we needed to look more widely at the total process.

I reminded the audience that ICE had published 'Our World and the Civil Engineer', that suggested that we were responsible for much of the essentials of modern life — the muscles and sinews which hold our society together (bridges, roads, railways, dams, airports, docks, tunnels); the provision and maintenance of its heart and lungs (clean water, natural resources in, waste out); transport for safe and effective movement; and energy to make it all work (offshore gas and oil, nuclear, hydro, tidal and wind power). We are deeply concerned with the whole environment of life — reducing pollution, responding to coastal erosion, restoring contaminated soil, disposing of hazardous waste.

There is an enormous programme to which we will make a massive contribution if our civil engineering is not too small and if we see ourselves as technologists addressing the 'why' questions as well as technicians addressing 'how' questions. We have to be at the forefront of the environmental movement rather than take up defensive positions. Yet the world remains desperate for new infrastructure provision as well as its efficient operation and maintenance.

My generation could make a particular contribution. The 21st century is for the next generation but the near retired, or already retired, have the time and perhaps the wisdom, and hopefully the energy, to help the younger members to chart the way forward. I emphasised help, because we could not do the job for them. It is their 21st century, not ours. Simply to plan and design infrastructure was not good enough. Ultimately implementation, efficiently and effectively, had to be our aim — getting the right things done well. For that reason we had to develop our management skills.

I was not seeking to become the Transport President of the Institution and my address would not concentrate on transport issues. Transport engineering was not *the* core specialisation of civil engineering, any more than structural engineering, water or ground engineering were. The extent to which transport engineering, and particularly transport planning, were properly matters of chartered professional expertise had been the subject of sometimes heated debate. I did believe however, that the study and practice of transport did have particular messages for civil engineering more widely. "It is quite impossible to be a successful transport engineer without paying due regard, for example, to the political dimension, good project management, the demand as well as the supply side of the equation, the human dimension, the environmental interface, and charging policy. All of those are more relevant to civil engineers than they have sometimes been regarded in the past. They will need to be central to our thinking and our practice well into the 21st century."

Thinking about the role of President it occurred to me that anything a President achieves during his term of office depends on what he has done in preparation during three previous years. Achievement also depends on developing a sense of continuity. A great Institution had to change to meet the challenges of now and the upcoming future. But change that zigzags is destructive and, furthermore, merely runs the Secretariat ragged.

The occasion of my Presidential Address provided a welcome opportunity to acknowledge three of my mentors who, together with others, had made such an important contribution to my career over the years.

I had formed a view of what my relationship with Council should be, having seen at least one predecessor fail to persuade Council to adopt a preferred policy. Having been instrumental in developing a new allocation of responsibilities at the top, I decided that proposals by the Executive

**PHOTO 11.1.　Mentors — Tom Beagley, Dept of Transport; Mrs Peter Stott, widow; Anthony Bull, London Transport.**

Committee should be presented by the Senior Vice-President (David Green), its Chairman. I would play the role of 'Speaker', so that members would not feel steam-rollered.

Of course, for 12 months the President is in a leadership role, although supported by the DG as 'Permanent Secretary', and that is what members expect. Previously the SVP had been Chairman of the Finance Committee for one year, while the President chaired the Executive Committee. My colleagues accepted a proposal to have an appropriate person to chair the Finance Committee for three years, and for the SVP to chair the Executive Committee for one. In this way the President was not trying to push his own

proposals through Council. This was also consistent with my view that anything that a President tries to launch after he has become the President is
already too late.

## 11.5  Future Voice for the Profession

A challenge arose during the presidential first few months — what to do with
the *New Civil Engineer*. It had become a financial burden on ICE, and took
up a disproportionate amount of time in Council. Moves had been set in
train to see whether a sale was a possibility, but this was confidential. I had
not been party to any negotiations, and was brought in immediately when a
decision was made to sell the NCE.

On the 10th anniversary of the sale to Emap, Antony Oliver wrote
(NCE, 2005),

> "'When President Tony Ridley took the unusual step of writing to all ICE
> Council members urging them to attend the 12 June 1995 meeting, they
> probably suspected something big was in the offing. But few foresaw that the
> debate would centre on the sale of the ICE's magazine publishing interests,
> and with them *New Civil Engineer,* the magazine of the ICE and voice of the
> profession. As it turned out, a packed Council meeting agreed the £6 million
> sale during an unprecedented 'closed session' — the room was cleared of
> secretariat and observers leaving only voting members in the chamber.'
>
> 'After a two hour debate, Council was asked to decide whether to do
> nothing and see the publishing business gradually slide into loss, invest more
> cash into the business to modernise it, or sell it and realise valuable cash to
> ease the ICE's financial woes. Pressure had been growing on the ICE for
> some time to realise cash from its publishing business, particularly since it
> was carrying a £4 million overdraft — a legacy of the £15 million revamp of
> the OGGS headquarters. But it was also aware that the publishing business
> needed substantial investment in technology to keep it moving forward —
> investment that could easily distract ICE from core activities.'
>
> 'Until that point only a select group at the ICE knew about the on-
> going negotiations. DG Roger Dobson and President Tony Ridley led the
> discussions — Ridley had been brought in after the premature death of
> President Edmund Hambly two months earlier. This duo was supported by
> VP David Green, Thomas Telford, Chairman Roger Sainsbury and FD Bill
> Cormie. All were bound by a strict confidentiality agreement. The reason

for secrecy was that two bidders were in the frame, one of which had no idea that it was not alone. The Builder Group, owner of *Building* magazine, had made the original running — unaware that Emap had opened parallel discussions with Dobson's team in secret. Both were bidding to buy the ICE's magazine publishing portfolio. This included the weeklies, NCE and loss-making *New Builder*, plus monthlies *Civil Engineering International, Ground Engineering, Water & Environmental Management, Highways and Offshore Engineer.* Both offered similar prices.'

'Council voted overwhelmingly to accept Emap's offer. Dobson described the deal as "a classic win–win" for Emap and the ICE and emphasised that it was "not a simple sale of the family silver" as it ensured continued magazine quality without exposing ICE to "unnecessary risk". President Tony Ridley agreed. "We are here to look after the well-being of members and develop the Institution to be fit for the 21$^{st}$ century", explained Ridley. "Council believes that this arrangement assists rather than detracts from it".'"

In the normal way a President does not get his photo on the front of NCE until about the time in November when he assumes the role. Mike Winney had decided that he would like to come to our cottage in Devon for the interview. So I appeared in the 20 July 1995 edition in walking gear, sitting on the Clapper Bridge at Postbridge on Dartmoor, with the title 'Ridley Sits In.' The six-page article (NCE, 1995) gave a rather good summary of my career to date, and included photographs of me travelling across the USA, on the model train in Dipwood House, and with Philip Essig at the time of Eurotunnel.

'Tony Ridley', said Winney under the heading, 'Transport to the Top',

"'is a big man. He talks authoritatively and loudly with a Wearside accent (really!), which might just have a transatlantic influence, while exuding the calm confidence of a preacher or head teacher. That confidence stems from spending many of his 61 years at the helm; as DG of the Tyne and Wear Metro, from its inception to the start of construction; MD of Hong Kong MTR, during construction of the initial system; and DG of London Underground, during the 1980s and its major modernisation. Following a turbulent spell with Eurotunnel Ridley returned in 1991 to his original calling in academia and became a professor at Imperial College.'

'In many publications and presentations Ridley has developed a clear thinking and philosophical approach to civil engineering and transport matters in particular, that is well matched to leadership of the profession at a time when engineers have been forced on the defensive about the environment. … His approach to problems, great or small, is to ask fundamental questions such as, what are we trying to achieve? and follow through to a logical answer. … (He) defines engineers as people who get things done; get things done well, but get the right things done.'"

This was accompanied by an editorial, 'Lead to Follow', in which Winney said,

"'This week's cover story about ICE President Tony Ridley should be an inspiration to all young engineers. Ridley's life story shows getting to the top can be realised by anyone. His career began, and is likely to finish, as an academic. In between Ridley has had his hands on the levers of power making things happen, creating two of the world's great metropolitan transport systems and running a third. Sheer hard work, persistence and an ability to recognise, then doggedly pursue opportunities, set Ridley on the path to taking his place as President. As a student he was not particularly outstanding, one of the crowd really, at a good engineering university where the staff featured few big names.'

'In retrospect what really marks Ridley out as different from his peers is something peculiarly appropriate for an engineer, a 'brass neck'. When career opportunities have appeared, Ridley has gone for them hard, putting everything else in second place. Having got to these positions of power he has justified the faith of those who had been persuaded by the brass neck. He has shown he is capable of running major projects and running them well. And the proof is that job offers have continued to come to him long after he passed 40. Graduates of 1995 please note and follow suit. There is a place for you as ICE President in 2035 or thereabouts.'"

What Winney said was flattering but, I hope, reasonably accurate. It certainly seemed to justify my putting my professional life story on paper, to offer advice to future generations.

## 11.6 Future Framework

Stuart Mustow was President of ICE from1993–1994 and influenced my thinking greatly. He was a lifelong local government man, until he 'retired' into consultancy. He had risen to become the County Surveyor of the West Midlands 1974–1986. His Who's Who recreations typically included 'church and social work'. He could have remained embedded in UK engineering, but it was Stuart who first lifted my ICE eyes to the wider world. In his review of his presidential year he said that,

> "The network of Local Associations is expanding into mainland Europe, North America, the Middle East, Hong Kong, SE Asia and Australasia. We have to move from our traditional assumptions that Local Associations are British phenomena to a position of seeing them as the norm worldwide wherever we have sufficient members to support them. Furthermore, they should be consulted about aspects of Institution policy in the same way as the Local Associations in Britain. ... We are increasingly an international Institution with a head office in London, rather than a UK institution alone."

He went on, discussing Organisation for Change,

> "Professional institutions are conservative bodies with organisations which are not designed for rapid changes in emphasis or direction. Despite significant strides in its corporate management in recent years and the respect in which its contribution to national debates is held, ICE remains rigidly hierarchical, slow to change and weak in strategic planning and policy development." (Mustow, 1994)

His contribution was recognised in the new format of the Annual Review (1994) which referred to the attitude survey that confirmed the need for ICE to take specific actions to secure greater recognition for the civil engineer's role in society at large. 'Mustow's Presidency to a great extent anticipated these, and the benefits of his presidency reminded us all that ICE has never stood still in proclaiming its members' contribution to the world's well-being.'

The Foreword to the Annual Report (1994) said, 'Civil engineering is becoming increasingly diverse. Civil engineers manage the construction, innovation, promotion, design, construction, operation, maintenance and eventual removal of the amenities of modern civilisation'. This obviously struck a chord with me. Then, in the Annual Review (1995) I was able to report that Council had agreed to commemorate the life and work of Edmund Hambly with a new Edmund Hambly Medal, to be awarded annually for the creative design of an engineering project that makes a substantial contribution to sustainable development, saying that it 'will be a lasting testament to his Presidency'.

In the Annual Review (1996) David Green said, 'As always, it is my privilege as current President to report on the major part of my predecessor's presidential year. Tony Ridley's 'year' was in fact 18 months, due to the tragic death of Edmund Hambly in March 1995. (His) Presidency is likely to be remembered as a chapter of critical self-appraisal within the civil engineering profession. It saw the launch of our Future Framework initiative in September (ICE, 1995) that led to the establishment of a Presidential Commission to propose the future role, nature, structure and constitution of the ICE. The stage for change was finally set with the publication in September 1996 of *Whither Civil Engineering?* — a major discussion document that puts the Future Framework report into the context of the future of the civil engineering profession as a whole. It argues that civil engineers should aim to provide a total infrastructure service as it moves into the new millennium. Chaired by David Cawthra, the Presidential Commission has consulted extensively using *Whither Civil Engineering?* the Future Framework and SAID reports as primary references.'

In discussing governance in 'The Civil Engineers', Ferguson and Chrimes (2011) say, 'In 1996, President Tony Ridley launched the most thorough review ever, deliberately choosing an 'outsider' to chair his commission. David Cawthra, former Chief Executive of Balfour Beatty, had little previous ICE experience — though he was elected a VP later that year. His comprehensive July 1997 report contained 228 recommendations which were to pave the way for the Institution's development in the 21st century.'

The Future Framework report (ICE, 1997) was presented to Council on 9 September 1997, by which time I had nearly completed my Past-President term, and most of its recommendations were accepted.

## 11.7 Inaugural Brunel International Lecture ('How Do We Advance Technology into the Third Millennium?')

Towards the end of 1998, a year before my retirement from Imperial, John Whitwell, Deputy DG of ICE, invited me to give the first of a new series of lectures, at first to be called the Telford Lecture but which became the Brunel International Lecture. This was a good opportunity to review and present my thoughts about Civil Engineering, and would give me the opportunity to describe the excellent work of Imperial's Civil Engineering Department. I threw myself into it, greatly assisted by a technician in the Department who developed a multi-media presentation. The name of the Lecture (Ridley T, 1999), subsequently published in an abridged form by RAEng (Ridley T, 2000), was changed to honour the Brunels, since Telford's name was already associated with the Telford Challenge, later known as Engineers Against Poverty. The lecture was given at ICE on 7 July 1999, and subsequently in Singapore, Hong Kong, Bangkok, Kuala Lumpur, Brisbane, Sydney, Washington DC, Toronto and Switzerland.

It began,

"It is a great honour for Imperial College and me to be invited by ICE to present the inaugural Brunel. Lecture. The invitation has encouraged me to draw upon both academia and practice and to use work from Imperial to think through the process of transferring civil engineering research into practice."

The announcement for the Lecture said that it would examine the legacies of Mark Isambard and Isambard Kingdom Brunel and look at the early history of civil engineering. It would reflect on the growing international stance of today's Institution of Civil Engineers, and would then address the distinction between technology-based science and science-based technology. It would draw on recent thinking about civil engineering R&D and the need for sustainability and would review current initiatives in the development of technology. The focus would then be on the means whereby good clients and good researchers could collaborate to advance research into practice, and would illustrate these ideas with examples taken from research at Imperial College. The examples described the work of no less than 12 colleagues.

It was extremely gratifying to have been able to give the Lecture, in all nine times, with additional requests from other parts of the world, thus carrying the name and reputation of ICE far and wide.

The lecture argued that the principle issues we had to address, if we were to advance civil engineering technology, were based on what civil engineering was about — practice, improving and maintaining, the built and natural environment, enhancing the quality of life in a sustainable way, and addressing the needs of society for both present and future generations.

> "'Technology is the study of technique, but it is also about products and processes. Civil engineering relies on science, but on technology-based science. In the late 20th and early 21st centuries biology and chemistry are increasingly important to the future of civil engineering, as are maths and physics. This reflects the broader, larger view of our profession that is appropriate for the future — but which our Charter already allows. The family of civil engineers now includes disciplines not traditionally thought to be part of the profession.'
>
> 'Technological change is a complex process that must be managed all the way from concept to the market place. Technological knowledge is cumulative and grows in path dependent ways. Technology has an impact when used in isolation. It may be implemented quickly but it often takes a long time.'
>
> 'Civil engineering education must recognise the importance of synthesis, of design, as well as more conventional analysis. Researchers would be well advised to address customer needs and market requirements. However, industry would be better served if it sought out good and relevant research more positively, if it developed more industry/academic partnerships.'"

A way ahead for both researchers and industrialists, was by asking in each case — what is the societal problem; what is the technological challenge; what is the business driver; how to define the research project; what are the findings (actual or potential); what are the potential applications; and, what is the mechanism (business process) for advancing research into practice?

> "'The process is iterative. The industrialist/businessman defines the problem. The technological challenge sets the research agenda. But the research equally defines the technological possibilities.'
>
> 'We are all prone to bemoan the fact that British scientists and engineers are enormously inventive but this inventiveness is not sufficiently often turned into commercially rewarding applications. The process is complex but there is a further issue relating to what some might call the 'academic profession'. Humans respond to various stimuli. We all seek to 'serve the customer' and look to stakeholders who are interested in our work. Who then is the 'customer' of the academic and what is the 'market'?

Certainly the students who come to be taught, as well as those who are guided through their research, are customers. So too are clients of research projects and of any associated consultancy. But part of the 'market' for academics are the journals to which they submit papers.'

'If we are to advance research into practice it is not enough for government, industry or Research Councils simply to sit in judgement on research proposals. They must actively seek out good researchers and, through mutual discussion, develop programmes that, in the case of civil engineering, address societal needs. I did not believe that the academic can be expected, except in rare cases, to be in the lead.'"

The vision of the future did not bear any resemblance to the swooping and whooshing elevated urban transit systems of the World Fairs of my youth. Those simply demonstrated how misguided forecasting physical things could turn out to be, and that are only a means to an end. It is services to meet the needs of society that civil engineers provide and it is creativity that is our essential contribution. The Latin 'ingenerare' means to create.

As we had said in 'Whither Civil Engineering?',

"To define a strategy, both a future vision and goal are prerequisites. Although it is impossible to predict precisely what the world will be like in the year 2020, I wrote in 1999, a vision should describe characteristics that we would want it to have. Those might be summarised as, a genuine improvement in quality of life for all, and long-term environmental, social and economic sustainability. The goal of Civil Engineering might then be to contribute towards achieving that vision, with its strategy focussing on the development of whatever structures, skills and technologies that are necessary in order to do so."

In closing it offered a challenge to the clients, consultants, contractors and academics who would take us on into the third millennium — to work together to develop research programmes that address the technological problems that had to be solved to provide business solutions that meet the needs of society.

## 11.8 Chartered Institute of Logistics and Transport

While in Hong Kong, I had been Vice-Chairman and a member of the Hong Kong Section Committee of CIT, and was subsequently invited to give the

Annual Overseas Lecture in London in 1980. I served 11 years as Council member of CITUK, and seven years as Chairman of the Transport Policy Committee.

The International Presidency of CIT began in October 1999. The Inauguration took place at the Annual Meeting in New Delhi. We had also just launched CIT World, the world-wide newsletter of CIT, another sign of growing independence from the UK. I said,

> "This is an important time in the history of our Institute. We have our own DG (Cyril Bleasedale), and I am delighted by the drive and enthusiasm with which he has taken on the role. We are no longer a child of the UK. We stand or fall by our own efforts. My first priority is to see the new CIT properly launched and our finances in good shape ... Next we must turn to communication. (We have) made a good start with the Newsletter, but we also need to develop web-based communications fit for the 20th century ... I applaud regionalism that complements internationalism, but regionalism as an alternative to internationalism is something I would deplore. Remember also that it is the international body that holds the power and responsibility to grant Chartered status to transport professionals, albeit acting through national bodies in many cases. But these are all organisational issues. It is the role of Council to ask what is it all for? While re-inventing ourselves we must re-examine our objectives and ask whether they are appropriate for our developing role."

Following meetings in Darwin and Birmingham, at an EGM in Kandy, it was announced that 84.7 percent of members voting (against 75 percent required) had approved a proposal that 'logistics' be embraced within the title Chartered Institute of Transport. The AGM had also approved the introduction, at the appropriate time, of the use of new post-nominals — FCILT, MCILT and AMCILT.

I stood down at my last Meeting in Singapore in October 2001, handing over to Jack So, the Chairman and CEO of HKMTRC.

## 11.9 The Universe of Engineering

In 1999 the then Science Minister, Lord Sainsbury, invited Bob Hawley, as Chairman of the Senate of the EC, 'to review the contribution the Council should make to add value to the engineering community, to the benefit of the UK Economy'. A review Group was set up, whose members were Sir Robert (Bob) Malpas as Chairman, myself and one other from EC, and two representing RAEng.

Our report (RAEng, 2000), 'The Universe of Engineering' was, to the best of my knowledge, the first to take the comprehensive view of the subject that we did. The title described the range of activities in which engineering is involved. We first produced a number of definitions, thought necessary because of the evident confusion in so many people's minds.

> *Science* is the body of, and quest for, fundamental knowledge and understanding of all things natural and man-made; their structure, properties, and how they behave. Pure science is concerned with extending knowledge for its own sake. Applied science extends this knowledge for a specific purpose.
>
> *Engineering* is the knowledge required, and the process applied, to conceive, design, make, build, operate, sustain, recycle or retire, something of sufficient technical content for a specified purpose; — a concept, a model, a product, a device, a process, a system, a technology.
>
> The *knowledge required* — know-what is the growing body of facts, experience and skills in science, engineering, and technology disciplines; coupled to an understanding of the fields of application.
>
> The *process applied* — know-how is the *creative process* that applies knowledge and experience to seek one or more technical solutions to meet a requirement, solve a problem, then to exercise informed judgement to implement the one that best meets constraints.
>
> *Technology* is an enabling package of knowledge, devices, systems, processes and other technologies created for a specific purpose. The word technology is used colloquially to describe a complete system, a capability, or a specific device.
>
> *Innovation* is the successful introduction of something new. In the context of the economy it relates to something of practical use that has significant technical content and achieves commercial success. In the context of society it relates to improvements in the quality of life. Innovation may be wholly new, such as the first cellular telephone, or a significantly better version of something that already exists.

The report sought to illuminate and make recommendations about the main issues at the heart of an unsatisfactory national situation — the central role of engineering in society and the economy is not evident to the public at large nor to the media in particular; the popular perception is generally confined to manufacturing and major building works; the engineering profession is considered by many, including unfortunately many young, as a somewhat dull, uncreative, activity wholly associated with the 'old economy'.

The report described the very many people — engineers, scientists, metallurgists, programmers and many others, who practise engineering in one form or another, to a greater or lesser degree, in the course of their professional activities. It was much larger than generally recognised. There were about 2 million people in the UK who called themselves engineers, about three-quarters of whom had a professional engineering qualification; and 600,000 engineers with qualifications at the Chartered/Incorporated level, 160,000 of whom were registered by the Engineering Council. There were no reliable figures even to estimate the numbers of people in a wider engineering community, who did not call themselves engineers, but who practised engineering in the course of their work.

In 2014 EC and RAEng (2014) produced an updated view which said that, 'fifteen years on from the original Universe of Engineering report, which highlighted the importance of engineering and engineers to the UK, many of its messages remain poorly addressed'. In reading the report I did wonder whether this might be because we engineers have still not defined the true nature of what we provide. Certainly the report strongly espouses the concept of breadth, though with emphasis on technological breadth. But it remains for us engineers to fully address the message that our purpose is to look beyond the physical things we create to the services our 'things' provide for the good of society.

## 11.10 LTA Lecture ('Benchmarking: Productivity and Service')

This lecture was given while on a visit to Singapore as a member of an international team advising the Minister of Transport (Ridley T, 2008). Again, the aim was to summarise my views about a very important topic with emphasis on services to society, rather than on physical outputs.

> "'In this paper I shall discuss productivity and service in metros and emphasize the importance of the concepts of service, efficiency and effectiveness.'

'My first productivity comparisons arose some 25 years ago when I was the Chairman of the Finance and Commerce Sub-Committee of the Metropolitan Railways Committee the International Union of Public Transport (UITP) and wrote a paper, with Hans Meyer of Hamburg, on productivity comparisons between metropolitan railways. For the paper we collected data from 26 metros in order to carry out our analyses. Using data over a five-year period (1977–1981) a series of graphs was plotted, showing the trends over the period and comparisons between undertakings, including passenger revenue/total cost, passenger km/train km, train km/total staff, and staff cost/total cost.

'Various cost information was also collected and analyzed. This did raise some doubts because of the necessary use of exchange rates. As well as being rather volatile over time, exchange rates may not reflect significant variations in purchasing power between different countries. Other means have since been developed to overcome these difficulties, notably 'purchasing power parity'.'

'The most detailed analysis was carried out on the data from Hamburg and London — the two authors' own undertakings. Clearly, different undertakings operate in quite different environments, which may make simple comparisons invalid. For instance the level of subsidy provided to the operator; the passenger markets in which they operate; depending on the form and structure of the area they serve — population, employment, car ownership; the nature of the labour market; and the extent of capital infrastructure already in place. We started with one basic statistic concerning the overall performance of the undertaking (passenger revenue/total cost) and then broke it down into its component parts. This made clear the important fact that an undertaking that performs 'well' on one criterion may perform much less well on another.'

'On the basis of admittedly imperfect data, we concluded that Hamburg was 63 percent less effective in its revenue earning capability per train-km than London — not surprising, of course, given the much higher fares levels in London. Hamburg, on the other hand was 51 percent more effective in the cost productivity of train running (train km/total cost) than was London. There were many more comparisons in the paper that gave insights into the respective performance of Hamburg and London. However the actual numbers themselves, and the rankings in any 'league tables' developed, are not the object of the exercise. The essential question is how we learn from the numbers, the questions that are stimulated by them and the self-examination engendered, leading to improved performance. These comparisons are much less a question of competing with others, rather of competition with oneself.'

'At that time I was MD of London Underground. So far as I was concerned the principal value of the work was to encourage a more searching examination of our own performance on the Underground which, being the first, dates back to 1863 and was originally built by a series of separate private companies. On a number of measures, Hamburg outperformed London. A fraternal visit to Hamburg followed and the lessons learned were built into the Underground strategic planning process. This focused a whole series of initiatives that involved changes in the company culture and culminated in changes in Board structure. The work with Hamburg also formed part of the basis of the development of the case to be made for more investment in the Underground.'

'Learning from others is an essential management tool for any successful business. This, of course, is the attitude of the Singapore government in inviting several 'experts', as an International Advisory Panel, to hold discussions with the LTA. Several years later, after my work with Meyer and after becoming an academic, I resumed my interest in the topic. By this time 'productivity comparisons' had given way to the term 'benchmarking', now widely used world-wide within a variety of disciplines and businesses.'

'What then is benchmarking? In 1989, an employee of the Xerox Corporation in the USA wrote a paper that defined the term 'benchmarking' as the search for industry best practices that lead to superior performance. Xerox had been in a fierce competitive battle with the Japanese, whose approach was gradually to improve performance by getting the most out of existing resources. This has been described by some, as 'making the assets sweat'. Xerox's earlier outstanding performance had been undermined and its market share and profits were declining. Analysis suggested that both their quality and productivity were falling behind to a frightening extent. Their reaction was to introduce a strategy of 'total quality', based on a study of best practice among their competitors. Other firms in other industries quickly followed suit.'

'In many industries today there is a focus on processes as well as on products, which had previously been the centre of managers' attention. Comparisons of business processes lie at the heart of benchmarking. The need for data collection and comparison can lead to an erroneous focus on the production of 'league tables' showing who is better than whom. In fact, the essence of benchmarking is to create new attitudes of mind that will lead to superior, or at least improved, performance. The proper question is not on 'how do we look?' but 'what shall we do?' Where any organization appears in a 'league table' will of course depend on managerial performance, but it will also depend on history and many other factors.'

'It is hardly surprising, for example, that Singapore MRT shows superior performance to London Underground when we consider the age of the Underground, and what the designers of MRT had learned from the experience of others over the previous century. The issue is — what lessons can London learn by examining processes in Singapore or, indeed any other city, and vice-versa? There is much that managers can learn, but it is also true that no amount of managerial excellence can overcome, for example, neglect of replacement investment over the years. It is also the case that, where a series of data is collected and analyzed, one organization may be 'better' than another on one measure but be 'worse' on a second. Furthermore, an organization with 'superior' performance might still have much to learn, or at least insights to gain, from an examination of the processes of others. This is particularly true where its assets are relatively new. Inspection of the performance of older undertakings, where replacement has been neglected, can provide salutary lessons.'

'Benchmarking 'clubs' most often include organizations from the same, or similar, industries. This is because the processes employed are likely to be similar and therefore easily comparable. However, lessons can also be learned across industries. South-West Airlines in the United States reputedly learned much from a study of rapid turnaround of complex equipment at pit stops in motor racing. In addition benchmarking may be carried out by comparisons within an organization. It is probably easier to arrange but obviously less likely to bring in the 'fresh air' of outside thinking.'

'Xerox's benchmarking process involved planning; analysis; integration and action (develop action plans, implement actions and monitor progress, recalculate benchmarks). This last step — recalculate benchmarks — is instructive, with its emphasis on constant improvement. Benchmarking is not a one-off action.'

'A fundamental prerequisite for successful benchmarking is commitment to organizational change. Indeed the management of change is the greatest challenge to managers at the beginning of the 21st century. The disciplines of project (or programme) management, and the associated management of risk, are now increasingly being applied far more widely in business than simply to support the management of discrete investments. Because the management of change requires dedicated commitment and leadership from the top, benchmarking should clearly be seen to have the support and drive of top management. Equally, because benchmarking regularly involves partnerships between organizations that may be competitors, often using confidential information, a high degree of trust is required. Thus the partners have to work out a formal method of working.'

'It is crucially important that managers understand three aspects of productivity performance, those that are under the control of the organization; are inherent in the cost structure of the industry; and those that pertain to the particular geographic and economic circumstances that a metro operates in. Given that transport is a highly politicized occupation, managers need to understand these distinctions if only in 'self-defence' against their masters.'

'My simplest example of this relates to a comparison of the financial fortunes of Hong Kong MTRC and Singapore MRTC. Both Hong Kong, where I held executive responsibility, and Singapore, where I was first an advisor and then an academic, have world-class metro systems. In Hong Kong revenue has covered operating, maintenance, renewal and construction costs. In Singapore construction costs were not covered. Does this mean that the people in Hong Kong (including myself) were somehow 'better' than the people in Singapore? Not at all. The circumstances are very different, particularly the urban densities. The point is that both Hong Kong and Singapore defined their own objectives in their own context and met them — but importantly, met them in an affordable way.'

'The main requirements for successful benchmarking are a strong commitment from top management to act on any major opportunities for improvement as they are revealed; a small amount of training and guidance for employees who will have to gather the information needed to identify and analyze best practice; and authorization for employees to spend some of their time on benchmarking activities. Of these, the most critical is top management commitment. To prevent benchmarking becoming an academic 'snapshot' of performance, senior management needs to own the process and be seen to be steering it.'

'Studies of benchmarking have discussed the role of Critical Success Factors (CSFs) and Key Performance Indicators (KPIs). To be successful it is essential that senior managers decide what is the mission of the organization. This is crucial to provide all employees with a clear idea of the objectives that are to be attained. It should therefore be communicated to everyone in the organization in clear and unambiguous language.'

'The engineer or the operator may focus on what they can do to make their system, or their project, productive or efficient. But this, on its own, is 'old engineering'. Today's engineers and operators recognize that the economic concept of supply and demand is essential to their task. Thus service to the customer becomes paramount. Other engineers may have to deal with the complexities of steel, concrete, water, electricity and the like. Those of us in the transport business have to deal with the most difficult 'material' of all — people.'

'We will not satisfy the customer unless we understand the nature of the service we provide and what the customer wants from that service. Public transport has been moving along a spectrum in recent years. Originally the ethos was that of production, but it moved on to that of the market, whereas the now the emphasis is on service.'

'What is a service? It is essentially a process and, while it is not physical in contrast to a product, service cannot be stored since it is produced and consumed at the same time. Another characteristic of a service is that the customer becomes part of the service process. It has taken some time for engineers, indeed many operators, to comprehend the crucial differences between production and service.'

'If, as I believe, the task of engineers, or of operators for that matter, is to 'get things done' then we need to look at other aspects or productivity and service. We not only need to get things done 'well' (be efficient), we also need to get the 'right' things done (be effective). There is a strong tendency to confuse efficiency and effectiveness. What we mean in our case is service efficiency (which is often the primary concern of the 'government customer') and service effectiveness (which is clearly of concern to our 'passenger customer'). There is no merit in efficient, but ineffective, public transport (because, for example, it is in the wrong place); nor in effective, but inefficient, public transport that costs too much and thus attracts low ridership.'

'If we are to focus on the customer then we must be both efficient and effective and provide a service to both the government customer and the passenger customer that is cost effective, in that it provides a good service that meets customer needs at an affordable price — in terms of fares or subsidy or both. Service efficiency may be measured by the level of service output relative to the resource cost of supply of inputs, while service effectiveness may be measured by the demand for use of the service relative to the level of service input. A cost effective service has attributes of both.'"

In summary this paper discussed the importance of focusing on the 'service' that metros provide to the public, rather than on technology and production. However, an understanding of performance, based on comparisons with others, can bring great benefits to organizations if they proceed in a measured way and, particularly, if they treat comparison as a self-learning process from which actions must follow if improvements are to be achieved.

## Bibliography

Annual Report (1994), ICE.

Annual Review (1994), ICE.

Annual Review (1995), ICE.

Annual Review (1996), ICE.

Beck B (1988), *Civil engineering — 2020, Department of Civil Engineering*, ICL (unpublished).

Ferguson H and M Chrimes (2011), *The Civil Engineers — The Story of the Institution of Civil Engineers and the People Who Made It*, ICE Publishing, Westminster, London.

ICE (1997), Future framework report.

Muspratt M A (1986), Civil engineering in crisis, in *Journal of Professional Issues in Engineering*, ASCE, Vol. 112, No.1, pp. 34–48.

Mustow S (1994), Presidential Year 1993/1994, ICE.

NCE (1995), Ridley sits in, 20 July.

NCE (2005), Sale benefitted both ICE and NCE, 16 June.

Philips J B (1961), *Your God is Too Small*, Macmillan Paperbacks, New York.

RAEng (2000), The universe of engineering, a UK perspective.

RAEng (2014), The universe of engineering, a call to action.

Ridley T (1994), W(h)ither civil engineering!, Discussion Paper, ULCTS, ICL.

Ridley T (1995c), Is our civil engineering too small? Presidential Address, ICE; (abridged), in *Proc. ICE*, Vol. 114, No.1, pp. 40–41.

Ridley T (1999), How do we advance civil engineering technology into the third millennium? Inaugural Brunel International Lecture, ICE, London.

Ridley T (2000), Civil engineering technology in the third millennium, in *Ingenia*, RAEng, London, Vol. 1, No. 3, pp. 53–57.

Ridley T (2008), Benchmarking: productivity and service. In *Journeys*, Land Transport Authority, Singapore, No. 1, pp. 34–43.

Ridley T *et al.* (1996), *Whither Civil Engineering?* Thomas Telford, London.

## Personal Correspondence

Freeman R to T Ridley (12 October 1996).

# 12

# Project Management

## 12.1 Risk and Collaboration with the Actuarial Profession

In 1994, as Senior Vice-President, ICE, I was asked to represent the Institution at a meeting about Risk and Finance for Capital Projects at the Queen Elizabeth II Conference Centre on 1 December, organised by the Institute of Actuaries. My first thoughts were — what do actuaries know about projects?

I had been teaching project management to third year undergraduates at Imperial, with a simple introduction to project risk. Somewhat puzzled, I went to the Conference (Institute & Faculty of Actuaries, 1994). The first speaker was Alistair Darling, then the Opposition Spokesman for City and Financial Services, on 'A View from the Labour Party'. The Private Sector Initiative was, he said, here to stay. But he went on to outline the 'substantial differences' between the government and Opposition as to the method by which decisions were reached.

The topic was clearly significant and the list of speakers impressive. For me the most significant speaker was Chris Lewin, Chairman of the Capital Projects Committee of the Actuarial Profession. He spoke on 'Capital Projects', briefly covering risk analysis, evaluation and management, and discount rates. He suggested that actuaries were already at the heart of some of the most important and exciting financial decisions that have to be taken by companies and by government, particularly where those decisions have long-term implications. An actuarial approach could assist in the analysis and management of risk.

He, and actuarial colleagues, had been researching the topic over more than two years. A CBI survey had found that only one-quarter of companies used quantitative methods to assess project risk and just over one-third used modern discounted cash flow as their primary method for appraising projects. It was clear that the actuaries were looking for new markets beyond their traditional involvement in life insurance. It seemed he was someone interesting to know. We went on to have a fruitful liaison over the next several years.

After a series of discussions we decided that the two professions should work together — civil engineers and actuaries (the Institute in London and the Faculty in Scotland, who later came together as the Institute and Faculty of Actuaries). We had a mutual interest, but different objectives. The actuaries were looking for new markets. I wanted to help to 'educate' my profession about the importance of risk, which became possible as an upcoming President. A Working Party was set up with a varying membership over the years, but Chris Lewin has remained centrally involved throughout.

Mike Nichols was persuaded to become involved and he put a great deal of work in, and used the resources of his firm to assist. The outputs of the early years were a first and second edition of *Risk Analysis and Management for Projects* (*RAMP*, 1998, 2002), which have not only had good sales, but also were commended by the Treasury. After the second edition we had turned our attention to strategic risk and, with the assistance of the Universities of Bath and Bristol supported by a grant from DTI, produced *Strategic Risk — a Guide for Directors* (*STRATrisk*, 2006). It said,

> "'Strategic risk consists of the most important risks that an organisation faces, i.e. possible future scenarios that would make a material difference (for better or worse) to its ability to achieve its main objectives, or even to survive. These risks differ in magnitude from project risks or operational risks, which are generally more limited in their impacts. Strategic risks are more strongly influenced by people's perceptions and their behaviour. They are more dynamic, uncertain and interconnected, and therefore they often need to be managed as complex processes rather than discrete events.'
>
> 'The (company) board needs to be fully involved in the risk management process, because it is usually only at board level that a sufficiently holistic approach can be taken, bringing together and prioritising the major

risks (and opportunities) in various parts of the business and making connections between them; the Board, working collectively, is likely to be multi-disciplinary and have longer and broader experience than most members of senior management; it is the Board which has the authority to ensure that risk management is given sufficient attention throughout the business, despite other pressures.'"

More recently a third edition of *RAMP* (2014) was published, with substantial enhancements to take account of the latest thinking about the management of uncertainty, clearer ways of presenting risk to decision-makers through simple scenarios and risk summaries, better criteria for determining whether risk responses should be adopted, and methods of coping with social and environmental risks in infrastructure projects. Although I was not an author of this edition, I was delighted also to write a dedication to Mike Nichols.

It is worth repeating here comments on good practices (Chapter 5)

"'Major projects pose a huge challenge to the project sponsor, given the importance, scope and time scale of the impacts and risks. Inevitably uncertainty is always present. Great care is needed, and a systematic project development process required, with risk analysis and management at its heart, to identify the right project and ensure its technical and stakeholder assessed success.'

'Practical project development requires the sponsor to identify the stakeholders early and thereby to form an understanding of who could influence the project. He needs to engage proactively with them and manage their expectations as the project is developed. This should be seen as additional to, and usually preceding, any formal requirement for public consultation. Effective project development requires balancing the sponsor's policy objectives with stakeholder views and agendas, so that the project is shown to have robust viability and strong stakeholder support stakeholder support.'"

## 12.2 Major Projects Association

MPA is a 'membership association for organisations engaged in the delivery and development of major projects, programmes and portfolios.' It was established in 1981 with the key objective of sharing experience, knowledge and ideas about major projects, both successes and failures, to help others to avoid mistakes and incorporate good practice so that future projects would be better initiated and delivered.

I became a Board Director in 1995 and stayed past the silver jubilee in 2006 until I stood down in 2009. On my departure the Chairman, Sir Robert Walmsley, said of me, "He has served on this Board since 1995 — making him the second longest serving Director in the history of this Association. Over his long and varied career it is possible that Tony has amassed more experience in major projects than every one of us in this room. He always brings this experience to bear upon our Board meetings in his own personal style — very 'respectfully' suggesting that perhaps the Chairman might wish to consider a different angle! Tony has contributed so much to this Association — wisdom, guidance and wit." I had served under other chairmen, in addition to Robert — Sir Michael Palliser (1994–1999), who had been Head of the Diplomatic Service, and Sir Alan Cockshaw (1998–2004).

## 12.3  Association for Project Management

In addition I held the Presidency APM 1999–2003. I was invited by Chairman Don Heath, a retired British Rail engineer who had modernised the East Coast Main Line rather successfully. My task was not onerous in terms of time involved. But it did have a somewhat dysfunctional Council that needed to be sorted out, in which I played a small part. It has happily come through that and is an increasingly significant and successful professional body. APM's history (APM, 2010) briefly records my professional career and said, 'he knew that APM had to change, if it was to move forward, and urged the association to "fix our processes".

## 12.4  Some Projects

As a professional who had many years of experience of project management, over the years I had several consulting/advisory jobs, mostly during my time at Imperial, sometimes to add experience to a consulting bid. I paid visits to Baghdad in the early 1980s to support work by a consortium of British consultants who developed a metro design for that capital city. I am convinced

Baghdad
Rapid Transit
Authority

هيئة مشروع النقل السريع

Draft Pre-Design Report
November 1981

Volume 1
Main Report

General Consultant
BMCG

**PHOTO 12.1. Baghdad, the Metro that Got Away.**

that if Saddam had gone ahead with the design that we produced for him, rather than spending a fortune on going to war with Iran instead, Iraq would have been in a much better shape today.

I also supported a proposal by British companies to design and build the first part of a new metro in Shanghai. The record shows that the Germans, assisted personally by Helmut Kohl, won the first contracts, and not the Brits.

Most of the experience was not in the stable environments of HK or Singapore, where projects had been identified after exhaustive analysis and decision-making processes were in place, but in developing countries where the challenge was to help identify good projects and establish such processes in often testing environments.

## 12.5  Bangkok

In July 1992, the World Bank discussed  their concern about Bangkok with me. They said that, following the government's decision to dispense with a proposal by the Canadian firm Lavalin, to develop a Skytrain for the city, I would shortly receive a call from a member of the Cabinet. He intended to invite me to visit Thailand to advise the government on 'what to do next'. He duly did so and said that he would like me to meet Dr Phisit Pakkasem, Secretary General of the National Economic and Social Development Board, who would be visiting Britain the following week, while attending the university graduation of his daughter. A meeting was arranged for 15 July at Dr Phisit's hotel, the Kensington Hilton.

Bangkok is by far Thailand's largest city. The national government had supported metro development ever since it was recommended by a German study in 1979, but there had been no progress. Meanwhile traffic had become terrible and faith in central government had waned. By the late 1980s Thailand was a leading proponent of pursuing private investment in infrastructure. Government's transport agencies were encouraged to find an international partner, to implement a megaproject via a concession. Many did and signed concession contracts but the projects often conflicted with or duplicated each other; and little happened in practice.

This log-jam was broken in 1991 when the Mayor of Bangkok, who had considerable clout as well as political legitimacy, gave up on the national government and advertised in the press inviting a proposal that would lead to a full build, operate, transfer (BOT) transit concession which would not require any government funding. This was the beginning of what was to become Bangkok's first metro — a fully elevated system developed by its concessionaire Tanayong that, after near-bankruptcy, has become a considerable success and is known as BTS or Skytrain.

In mid-1992, a political crisis led to the King inviting a widely respected technocrat, Anand Panyarachun, to form a technocratic government, with a short time-frame, to restore credibility to the country's leadership. He perceived the need to develop a government sponsored metro project to re-establish its competence in infrastructure development. The Lavalin Skytrain concession was cancelled. I was then invited to assist the government develop its own project for early implementation. I immediately asked Roger Allport of HFA to assist with this work. After retiring early from Halcrow he joined RTSC, on a part time basis and completed a 'mature' PhD (Allport, 2011).

Allport and I had been involved together when he was part of the Halcrow team that carried out the study for the Singapore NE corridor rail extension. Halcrow's proposal had included the establishment of a Peer Review Team, to be associated with the work, consisting of Sir Alan Muir Wood and David Buckley of Halcrow, and myself of HFA and London Transport.

It is a feature of a consultancy that I have always believed in. The reviewers have two functions — to cross-examine the working team about their work before they present to the client, and to reassure and to be able to answer questions by the client during the course of discussion after their presentation. The experience of the reviewers is nearly always appreciated by clients in these circumstances.

In our meeting with Dr Phisit we learned what he thought the problems to be solved were. Two days later we sent a proposal (Ridley T, 1992). The accompanying letter said, 'I understand that you are seeking my urgent assistance to find a way ahead following the recent Skytrain decision. You have emphasised the need to examine important institutional questions so that a Cabinet committee might take far-reaching decisions before the end of September. However, important funding questions have also been mentioned as well as issues such as potential conflicts between the three different transit proposals, alignment, and whether there might be a case for considering underground construction. ... I am proposing the assembly of a number of people from a variety of backgrounds and affiliations. I am sure that you will appreciate that, if I am to give the advice which I understand that you urgently need, it is crucial that a decision to proceed is given quickly' (Ridley T to Phisit, 17 July 1992).

Phisit soon replied, 'First, the Thai government has approved your proposal concerning mass transit in Bangkok in its cabinet meeting on 21 July, Second, the Thai side has set up a working group to come up with more detailed terms of reference. Your input in finalising the TOR during

28–30 July will be appreciated. Thirdly, the Office of NESDB is setting up for you a schedule of meetings with key personnel in the mass transit area, including an appointment with the Prime Minister. Fourthly, kindly advise us as soon as possible about your exact date of arrival and flight number' (Phisit to T Ridley, 22 July 1992).

The next day I faxed back, 'I am pleased to hear that the study will proceed. The questions I shall wish to discuss and settle during the visit next week are the contract; organisational arrangements; and institutional questions. I am pleased to hear that a series of meetings has been arranged (Ridley T to Phisit, 23 July 1992). Fast work by any standards. Back in London I wrote, 'There is a great deal to be done in a very short time — effectively five weeks study, one week of review and two weeks to finalise my report. My approach will be to develop a Core Network which is capable of rapid implementation. This I interpret as avoiding controversy and major conflicts between Bangkok's megaprojects, focussing on an initial network 'anchored' at the Skytrain depot site and comprising some adaptation of the Skytrain and Tanayong networks. Initially, it will be my intention to develop the Core Network irrespective of sponsor and presuming a common technology, then to establish how it can best be divided between Tanayong and the new Metropolitan Rapid Transit Authority, and how the MRTA network can be implemented (Ridley T to Phisit, 30 July 1992).

The Team, admirably led by Allport, comprised professionals from several companies all with up-to-date experience of Bangkok's transport scene. It got to work and on 7 September I had returned to Bangkok to present the report to a Cabinet sub-committee. On 25 September the report, Bangkok Advice on Rationalisation of Rapid Transit Systems (Ridley T *et al.*, 1992), was presented to Dr Phisit under an accompanying letter. 'My report contains the recommendations made to the Cabinet Sub-Committee on 7 September and approved the following day, together with the basis for the Recommendations. I would add two points. First, while much of the work undertaken was of necessity preliminary and needs to be followed up, I am confident that the recommendations are robust. You have asked me to recommend the immediate next steps and these include further work that will produce estimates in which greater confidence can be placed. But they will not change the strategy that you have chosen. Secondly, having spoken to both Hopewell and Tanayong, I am in no doubt that we have been an important catalyst in developing an integrated rapid transit network for Bangkok from which all partners will benefit' (Ridley T to Phisit, 25 September 1992).

I next wrote to Dr Phisit, 'I have heard some good news and some not-so-good news.' I was pleased that a contract had been agreed with the MRTA, though it focussed on engineering and control. My concern was that Studies that I had recommended had been excluded from the contract. They were necessary for a number of reasons.

"'They define the system (station, line) capacity needs. Without them engineers will, for safety reasons and reasons of prudency, need to make 'conservative' (i.e. costly) design assumptions about possible ridership levels and station interchange requirements. These parameters have so far not been determined with the necessary degree of confidence — because virtually no state-of-art planning has been undertaken in Bangkok in more than 10 years. The cost of the necessary Studies will be offset many times by savings in the conservative design assumptions which otherwise will need to be made. My experience in Hong Kong, Singapore and elsewhere leave me in no doubt as to their worth.'

'The Studies define the financial characteristics of the Initial System. Only when these are determined can the MRTA's commercial brief be defined, to provide the cutting edge in management decision-making; the MRTA management be provided with the tools to ensure, through their day-to-day decisions, that the finances of the project are controlled — this is of crucial importance since the financial outturn is, to a very large extent, determined by the decisions MRTA take every day from now through to operations; finally, and importantly, the feasible funding structure for the project be defined so that funding sources can be lined up. The scale of necessary government equity can only then be confirmed and desirable debt structure determined.'

'The Studies, to a considerable extent determine — or constrain — the sources of possible borrowing for the project. At its simplest, the absence of Studies carried out by consultants of international repute will remove completely the possibility of funding by the major development banks, including the World Bank and the Asian Development Bank (ADB). As I said at the time of my first proposal and in my Recommendations to Cabinet, there has been almost no state-of-art transit planning in Bangkok and this is now essential. The Studies will substantially reduce the cost of the project and will not change the programme to implementation, providing they go ahead rapidly. They will vastly improve the operating and financial performance of the system, give MRTA a demanding but realistic commercial brief, and ensure that all desirable funding sources remain open. They will, in short, enable the Initial System to be the operating and financial success which throughout has been my, and I know your and the MRTA Board's, intent'." (Ridley T to Phisit, 19 October 1992)

Shortly afterwards the National Economic and Social Development Board proposed, and the Cabinet endorsed, the 19 km, 46,000 million baht, project. A newspaper report said that,

"It also endorsed the appointment of the MRTA Committee, chaired by the Prime Minister, to ensure speedy implementation of the project. The NESDB made the proposal on the recommendations of Professor Tony Ridley of the University of London, who was hired to study the rapid transit system after the contract of Lavalin International Group was nullified. The Government would fund about 70 percent of the project (30,000 million baht), the remainder being sought through loans from overseas."

November 1992 was effectively the end of my involvement, but it was another 12 years and several setbacks before the Blue Line entered into operations. Several major problems arose. The Thai Cabinet decided without warning to ban elevated infrastructure construction in the central 25 km² of Bangkok on environmental grounds. The partly elevated alignment I recommended had to be fully undergrounded, causing delay, large cost escalation and a legacy of high operating costs. There have been deep-seated institutional problems that have frustrated integration of the two metro systems and their integration with buses. And there has been uncertainty about future system expansion that impacts the Blue Line concessionaire substantially.

## 12.6 Jakarta

I have majored above on my relationship with people at the top in Bangkok, supported by written advice that attempted to have them understand just what was necessary to create a successful metro, and what a successful client needs to do. Today, much later than was hoped, Bangkok has developed its metro system. Jakarta has not yet met with the same success. I describe below some of the complexities which hindered the project.

On 23 November 1993, His Excellency Professor Doctor Ing B J Habibie, State Minister for Research and Technology of the Republic of Indonesia visited me in my office at Imperial College. He was on a wider 3-day visit to the UK, meeting members of the government and leading members of the engineering profession.

Habibie, who subsequently became both Vice-President and then President of Indonesia, was accompanied on his visit by John Coplin, seconded from Rolls Royce to be his UK Advisor. Habibie had shrewdly appointed five such advisors, from the UK, USA, Japan, Germany and France, who Habibie called upon whenever he wanted anything from those countries, or to visit them. Within 15 minutes he had, figuratively, put two first-class air tickets (one for Jane) in my hand and persuaded me to fly to Jakarta, all expenses paid but no fee, and to report to him what steps he should take with some transport issues — in particular, a fixed rail system through the middle of the capital. He also presented me with a copy of President Soeharto's autobiography (1991) wherein I first learned of Pancasila, the philosophical foundation of the then Indonesian state

On returning to the UK I wrote to Habibie with a short report, the object of the visit having been to determine in what way I might helpfully fulfil the role of Advisor to Minister Habibie (Ridley T to Habibie, 26 January 1994).

My reflections on the visit included Perumka (Indonesian Railways) and Academic contacts, but focussed on Urban Mass Transit. There were undoubtedly a significant number of well-qualified and competent government officials at the working level, who had done a good job in developing a preliminary programme for the urban mass transit system. This was based on a series of consultant studies previously carried out for the government of Indonesia that had somewhat variously recommended a metro, upgrading suburban rail and a bus rapid transit system as the solution to Jakarta's transit woes. It was clear that there was not yet unanimity about the most appropriate next steps.

A large number of important decisions remained to be made before an agreed system and construction programme could be started. In part this was because the apparent consensus achieved in the creation of a Consolidated Network was not reflected in the views of very senior members of government, who held a variety of views about what should be done and how the process should develop. This was not surprising and was certainly not unique to Jakarta and Indonesia. The essential challenge of the creation of an urban mass transit system, is the management of profound complexity — bringing together questions about the physical system, operational and communications matters, finance and funding, as well as organisational and environmental matters.

In an urban setting each of these is difficult in its own right, but it is the inter-relationship between them that presents the greatest challenge. Technical matters cannot be settled outside the context of political decision-making. Equally, political decisions have implications for technical, financial and other matters. Thus, in the development of an urban mass transit system, an unusually high degree of good communication between the technical and political levels of decision-making is necessary.

A new Authority was to be created by Presidential Decree. This would create the mechanism for taking the system forward. Indonesia already had experience of carrying out large, expensive and complex projects, but this experience did not extend to urban mass transit projects with the special problems they brought. The team that developed the Consolidated Network had already produced a Preparation Plan for further work. It was my understanding that this work would not now proceed until after the creation of the new Authority. It was also not clear how this work would be funded.

An early requirement was to obtain an indication of the financial feasibility of the plans. This would not create a firm Financial Plan which properly comes at the end of the planning process. However, there was much continuing discussion of underground versus elevated construction, heavy versus light rail, private versus public sector funding. None of these could be properly discussed, let alone decided, unless some indication of the financial implications was available. It was also very important, given the international interest in the projects, that government arm itself with adequate information for any negotiations with international consortia and/or governments. However much private sector investment, overseas export credit loans and International Bank funding was available, it was essential that GoI remain in overall charge of the development of the system. There were unhappy examples in other countries where this had not happened, which had led to both confusion and delay.

After this, future visits followed a regular pattern — fly through Singapore to Jakarta, overnight Saturday; relax in my hotel on Sunday, with dinner at the Coplin's in the evening; work four days with the Team; debrief to Habibie on Friday; return to UK overnight Friday. First of course, I had to have an appointment, and to assemble a Team. This took a long time, after submitting a proposal (24 February 1994).

In October, I wrote (Ridley T to Habibie, 11 October 1994) to let him know the latest thinking about the project in the capacity as his independent Advisor. I had paid a brief visit to Jakarta in August and Allport had spent one week there more recently. As a result of Habibie's approach to the UK government my advice was being funded by the DTI. Both Allport and I had engaged in conversations with officials within BPPT and other government departments. We were in close contact with the UK Embassy, and with John Coplin. It was my judgement that all of these relationships were working well. Much progress was being made, but a very large amount of work remained to be done to ensure the success of the initiative.

Eventually, after my position was confirmed, I visited Jakarta again, after which I wrote (Ridley T to Habibie, 14 December 1994) that there were some important decisions about the project to be made shortly. Government should decide how to handle 'competition' between potential private sector partners; decide the status of the alternative double-deck, toll road/transit proposals; ensure all GoI departments were committed to the decisions; examine, in detail, the consequences of different portions of underground and elevated system; examine, in detail, the consequences of the need for substantial GoI borrowing for the project; and establish a single, high-level government decision-making body for the project.'

I wrote again after a further visit (Ridley T to Habibie, 5 April 1995), and said that he had a great deal of international interest in the MRT project, but the world was not yet convinced that it would proceed and had top priority with senior members of government. Thus, it was essential to establish the Project Management Unit (PMU) and its Chairman, at the earliest opportunity; and to establish a Steering Committee which could meet regularly and frequently enough to make urgent policy decisions, of which there would be many.

This work could be funded by GoI or with the help of private partners. A decision must be reached quickly. Whichever path was chosen, government must participate in all parts of the process. However the Basic Design was funded and carried out it was important that he, as Chairman of the Steering Committee, should recognise the need regularly to receive advice from the PMU on policy decisions to be taken. The Basic Design was an iterative process and senior decision makers had to be involved.

A constant theme, both face-to-face and in writing was 'decisions, decisions, decisions'. I tried to explain how Haddon-Cave had done it, as Chairman of the HKMTPA. Habibie nodded sagely during each of the 'Friday audiences' but no process emerged. I tried a different tack, seeking to persuade the very bright young Indonesian professionals who were in the 'team' that they must advise Habibie about the decisions that had to be taken. They were aghast, "Oh no, we must wait until the Minister has decided what we must do." "But how will he know what to decide if you do not advise him, present him with alternatives?" The much-travelled Ridley had clearly not understood their culture. Then I began to realise that perhaps it was an impossible task.

Haddon-Cave was the Financial Secretary of a city-state of some 5 million people. Habibie was the right-hand man of a military President of a country with a population nearing 200 million, with countless islands and several religions. Habibie was chairman of a vast number of government industries, was trying to create an aircraft industry, and was very active in the Islamic community. He simply couldn't find the time. But if he could not find the time, was the whole concept of a successful Jakarta metro, with all its complexities, a non-starter? I never solved that conundrum.

After the next visit I wrote (Ridley T to Habibie, 12 September 1995) that he had asked me to tell him what he must do. I said that he had to find time in his extraordinarily busy diary to give sustained and regular attention to the project. Because the Basic Design process of reaching optimality was iterative it was essential that there be regular discussion (not just passage of information) and decision-making between the PMU and senior members of government. Somehow the opportunity needed to be found for himself and the Governor of Jakarta to meet with his working team not less frequently than once per month.

In the Summer of 1996, my post-visit letter (Ridley T to Habibie, 15 August 1996) said that lack of a number of major decisions was now putting the project timetable at severe risk. The siting of the depot was an example of this. … The conflict between a budget limit and the decision to put the whole system underground was one of the issues that had to be resolved. … The project would require the collaboration and agreement of many arms of government. Without this the project could not succeed.

After what turned out to be a penultimate visit, I was again stressing that the project had reached a crucial stage (Ridley T to Habibie, 17 January 1997). In March one of Habibie's lieutenants had written to the UK Embassy in Jakarta stressing my 'valuable help' and asking for its continuation. However, early in 1998, the DTI wrote to me saying that, given the dire state of Indonesia's economic prospects, it was premature to consider committing financial resources until the situation became more settled', with which I concurred.

I never returned, and the last time I met now-Vice President Habibie was when I was invited to join sundry captains of industry who were hosting him for a breakfast meeting during one of his regular visits to London.

However, times can change, as the following announcement shows. 'After 24 years of dreaming to have an MRT, some may even have forgotten it. Finally the dreams of Jakarta residents will come true', said the Railway Gazette (15 October 2013). Following economic challenges by 2013 Indonesia had re-emerged as a leading economy with the capacity to undertake and fund major infrastructure projects. Its first metro, that had its ground-breaking in late 2013, follows much of the advice I had given, in particular controlling costs by undergrounding only when necessary, building a system that could be extended, and identifying attractive financing. A soft loan has been secured from Japan on very attractive terms, and Japanese companies are charged with the implementation of the project's 16 km system.

## Bibliography

Allport R J (2011), *Planning Major Projects*, Thomas Telford, London.
APM (2010), A history of the Association for Project Management.
Institute & Faculty of Actuaries (1994), Capital projects — risk and finance, Conference papers, December.
Jakarta Proposal to the Indonesian Govt (24 February 1994).
Railway Gazette (15 October 2013).

RAMP (1998, 2002 and 2014), *Risk Analysis and Management for Projects*, Thomas Telford, London.

Ridley T (1992), Mass transit in Bangkok — steps to implementation, A Proposal for Advice (unpublished), 17 July.

Ridley T *et al.* (1992), Bangkok advice on rationalisation of rapid transit systems, Report to Government of Thailand, 7 September.

Soeharto (1991), *My Thoughts, Words and Deeds*, An Autobiography, Citra Lamtoro Gung Persada, Jakarta.

STRATrisk (2006), *Strategic Risk — Guide for Directors*, Thomas Telford, London.

## Personal Correspondence

Phisit to T Ridley (22 July 1992).
Ridley T to Phisit (17 July 1992).
Ridley T to Phisit (23 July 1992).
Ridley T to Phisit (30 July 1992).
Ridley T to Phisit (25 September 1992).
Ridley T to Phisit (19 October 1992).
Ridley T to B Habibie (26 January 1994).
Ridley T to B Habibie (24 February 1994).
Ridley T to B Habibie (11 October 1994).
Ridley T to B Habibie (14 December 1994).
Ridley T to B Habibie (5 April 1995).
Ridley T to B Habibie (12 September 1995).
Ridley T to B Habibie (15 August 1996).
Ridley T to B Habibie (17 January 1997).

# 13

# International

## 13.1 Commonwealth Engineers Council

CEC was founded in 1946, after a suggestion by ICE to IEE and IMechE, who were joined in a meeting by the national engineering institutions of Australia, Canada, India, New Zealand and South Africa. Over the years it grew in membership to a total of 18 in 1980 and later to some 50 countries. The Secretariat for CEC has been provided by the DG of ICE, though he has often delegated his responsibilities to a member of staff. In 1993 neither the President nor the Senior VP were able to attend a meeting of the CEC in Jamaica, so I was asked to stand in.

Two things stand out in my memory — meeting David Thom (New Zealand) and secondly Dato Lee Yee Cheong (Malaysia). Thom was an engineer who, in later life, had become an international environmental champion together with others such as the American, Don Roberts. Thom was the leading light among engineers worldwide seeking to make engineers more environmentally conscious. His paper was 'the New Technical Culture' (Thom, 1993). It was a pleasure to receive from him a large compendium, at about the time of his 90$^{th}$ birthday, which was the publication of his many papers 1970–2000 (Thom, 2014). He was kind enough to include me in a dedication to a small number of friends from around the world, 'who have provided both support and inspiration.'

Dato Lee was a Chinese Malaysian electrical engineer who became the President of CEC and then, with my support, President of WFEO (2003–2005). In his autobiography (Lee, 2010) says that, after discussions about the future of CEC, 'the Engineering Council and ICE UK decided that the UK would continue to host the secretariat. However, they were not prepared to support me until ICE VP Tony Ridley looked me over at the CEC Council meeting in Jamaica 1–2 November 1993.' Not true, but I did support him, then and later.

The 50th Anniversary of CEC occurred on 1996 and Lee encouraged the idea of a 50th Anniversary Conference, to be held at ICE Headquarters at OGGS. This was preceded on 18 March by an Inaugural Banquet and a Reception where the Guest of Honour, Her Majesty the Queen, met the delegates. Lee (President of CEC) and I (President of ICE) and our wives, jointly welcomed the Royal Party to ICE after which Lee escorted the Queen to meet the 250 delegates, from around the Commonwealth, together with Bud Carrol (USA) and Conrad Bauer (Argentina) representing WFEO, and Jose Medem (Spain).

When the Queen arrived she looked sad. She had just returned from Dunblane in Scotland where she had been to comfort the community which had suffered so grievously from the murder of some dozen children at the

**PHOTO 13.1. HM the Queen at ICE for the Commonwealth Engineers Council 50th Anniversary.**

hand of a single assassin. But, by the time she left us, she was glowing. There is no doubt that, not only are Commonwealth people enormously warmed by the Queen, she too is equally warmed by them.

The next day I received a letter from Simon Gimson at the Palace,

"I am sure that it was clear to you that the Queen had a marvellous time at Great George Street last night. Her Majesty wanted to be sure that I wrote to you as soon as possible to say as much, and to extend her thanks to you and the Institution of Civil Engineers for being such charming co-hosts with the Commonwealth Engineers Council. In view of the founding role which ICE took in 1946 on the formation of the CEC, it seemed most appropriate that the CEC 50th anniversary was celebrated in your splendid Institution's building.

Her Majesty was pleased to be part of the celebrations, and to meet so many engineers representing such a wide variety of disciplines from throughout the Commonwealth. This letter, then, comes to you and to the Institution with the thanks and good wishes of the Queen. Together with those, may I also add thanks of the three of us fortunate enough to be in attendance, for what proved to be a most convivial and stimulating evening." (Gimson S to T Ridley, 19 March 1996)

The Conference title was 'Engineering to Survive — Global Solutions for Sustainable Development' (CEC, 1996). It covered a wide range of topics — Threats and Opportunities, Water and Waste, Food and Agriculture, Planning and Legislation, Resources and Development. Many speakers covered a wide range of engineering disciplines.

In 2000, I followed Lee as President of CEC, which also introduced me to the World Federation of Engineering Organisations, widening connections and understanding of issues faced outside the circle of UK engineering.

## 13.2 World Federation of Engineering Organisations

One action by the UK Engineering Council, when I was a member, annoyed me considerably — a proposal that it should withdraw UK's membership from WFEO. Certainly the EC was wise to look to its policy and financial priorities, but it seemed to me that to withdraw would send a signal that it thought that UK engineering ended at the White Cliffs of Dover.

WFEO had been founded in 1968, by CEC among others, during which time the UK had been active participants. In September 1998 a letter was sent by EC to all its Nominated Bodies saying that ICE had asked the EC to

> "reconsider and, indeed, to put resources into trying to improve WFEO's effectiveness. At the same time Senate consideration of the strategy review has pointed to a need to incorporate a more thorough analysis of the EC's role in the international field — perhaps including consideration of delegating more functions to individual Institutions."

At least one body gave the rather terse response that,

> "if the ICE alone is so keen on retaining links with WFEO, for whatever reason (whether on account of the profession as a whole or principally its own members), I suggest that the EC invites it to take the lead forthwith on behalf of the profession, in line with Tony Ridley's proposal, and pick up the bill!"

And that is what did happen. ICE Past President Prof Roy Severn wrote to me in February 1999 suggesting that Jose Medem, the Spanish President of WFEO, would be prepared to travel to London to meet me 'so that he could exercise his undoubted charm and persuasion in WFEO's favour'. In the event the meeting was organised by John McKenzie, former DG of ICE and previously Executive Director of WFEO, at his club in London — the Athenaeum. On the same occasion Jose, on behalf of the Spanish Engineering Organisation, invited me to attend the next WFEO General Assembly in Madrid in November 1999, as a guest speaker.

The eventual outcome was a decision in July 2000 that ICE would join WFEO, until such time that the wider UK engineering profession decided to retake comprehensive representation — if ever. WFEO offered a substantially reduced subscription for the first two years. It was agreed that, initially, ICE as National Member would be represented by me in my capacity as the President of CEC, which role gave me a seat on both the Executive Committee and General Assembly. The formal ratification was to take place at the next General Assembly in Moscow in September 2001. Thereafter, I regularly attended Executive Committee and GA meetings until I stood down from CEC in Kuwait in 2009. As CEC President I was pleased to be able to help to promote the successful presidential candidacies of Lee Yee Cheong (2003–2005) of Malaysia and Barry Grear (2007–2009) of Australia.

## 13.3 United Nations

In June 1972, the first ever UN conference on human development had been held in Stockholm. Twenty years later the UN Conference on Environment and Development, the Earth Summit, was held in Rio de Janeiro. This set a series of activities in motion. One of them was the creation of the Commission on Sustainable Development, a body of more than 50 member states responsible for reviewing progress in the implementation of Agenda 21 and the Rio Declaration. It met annually at UN headquarters in New York.

Associated with the CSD were so-called multi-stakeholder meetings. CSD-9 took place in April 2001. Prior to this they had comprised Business and Industry (International Chamber of Commerce, World Business Council for Sustainable Development and World Energy Council; Women; Children and Youth); Indigenous Peoples; Workers and Trade Unions; Local Authorities (International Council for Local Environmental Initiatives; Non-governmental Organisations; and Farmers).

In 2001, for the first time, representatives of the Scientific and Technological (S&T) Community were invited to attend, namely the International Council of Scientific Unions (ICSU). ICSU, which comprised 95 national scientific members, the Royal Society among many others, invited WFEO to participate as part of their delegation. I had become a member of the Executive Council of WFEO by virtue of my presidency of the CEC in 2000. The topics were energy and transport. I contributed to the transport paper submitted by S&T, which relied heavily on 'A vision for transport 2020' (Ridley T *et al.*, 1997), prepared previously for the UK Engineering Council by ICE.

In January 2002, I again went with Lee, and others, to participate in the preparation for the World Summit on Sustainable Development to be held in Johannesburg in August 2002. As part of WSSD, the South African government organised a Forum on Science, Technology and Innovation for Sustainable Development. ICSU, WFEO and the Third World Academy of Sciences agreed to support the coordination of the Forum. Within the Forum, WFEO and the Engineering Council of South Africa organised a one-day session on 'Engineering and Technology for Sustainable Development'. The main objective was to highlight the crucial role of engineering and engineers in tackling sustainable development and poverty alleviation.

UN Secretary-General Kofi Annan had said that, "ten years after (Rio) a major obstacle to sustainable development for much of the world remains the lack of scientific and technical capacity ... the resources that have been made available for capacity building have been relatively meagre in relation to the growing requirements of developing countries. More resources are needed to support international efforts at capacity building".

At the Forum, there was a wide variety of papers about capacity building. Professor Juma of Harvard University, in addressing technology capacity building suggested that perceived lack of progress since Rio was as much to do with *lack of adequate scientific and technological investment* as with lack of political will.

In its 2/9 January edition NCE (2003) ran a piece headed 'Council puts sustainability top of education curriculum'. 'Past President Tony Ridley, who instigated the debate through his paper outlining his experiences at last August's WSSD, said,

> "A 21$^{st}$ century engineer is not the one I was educated to be in the 20$^{th}$ century. ICE has partly caught up, but still has a distance to go. Sustainable development is now absolutely central to Civil Engineering, and we must organise ourselves accordingly."
>
> "In addition to a revamp of engineering education, Ridley called for an immediate audit of the activities of ICE's engineering boards to ensure that ICE is fulfilling its role as a centre of knowledge and best practice. This view was backed by Council member and Environment and Sustainability Board Chairman John Ekins."
>
> "Civil engineers are still seen by many as uncaring philistines, and we have got to get a grip on this. As chair of the E&SB we have to get to the point where we don't need an E&SB, as it should be at the heart of all boards."

## 13.4 UN Task Force 10

The UN Millennium Project was commissioned by the Secretary-General, Kofi Annan, to advise on the implementation of the Millennium Development

Goals arising from the Millennium Summit in 2000. Ten task forces were established to address inter alia, hunger, education and gender equality, child and maternal health, diseases, environmental sustainability, water and sanitation, the lives of slum dwellers, and trade. Task Force 10, 'Innovation — applying knowledge in development', was deliberately and necessarily cross-cutting.

The task force coordinators were Calestous Juma and Lee Yee Cheong. I had met Juma when I attended the WSSD in Johannesburg in 2002. He knew Annan well, having worked for him at the UN as Executive Secretary of the UN Convention on Biological Diversity, and was by now a professor in the Kennedy School of Government at Harvard. He is also FRS and Hon FREng, not bad for someone who was born in a small village in Northwest Kenya.

The task force had a widely drawn membership, with representatives from UNCTAD, UNEP, UNDP, UNESCO and UNIDO, academics from Hong Kong, Kenya, Sri Lanka, UK and Uruguay and others. Everyone contributed to the whole document, and also made particular contributions based on our experience, mine being to Chapter 5 — 'Adequate infrastructure services as a foundation for technology'.

The Executive Summary to our report (Juma and Lee, 2005) began,

"Since their adoption at the UN Millennium summit in 2000, the MDGs have become the international standard of reference for measuring and tracking improvements in the human condition in developing countries. The Goals are backed by a political mandate agreed to by the leaders of all UN member states. They offer a comprehensive and multi-dimensional development framework and set clear quantifiable targets to be achieved by 2015. Meeting the Goals will require a substantial reorientation of development policies to focus on key sources of economic growth, including those associated with the use of new and established scientific and technological knowledge and related institutional adjustments. Countries will need to recognise the benefits from advances in science and technology and develop strategies to harness the explosion of new knowledge."

In all of the work of my later professional life, which was directed at assisting developing counties, I was reminded that the messages were equally relevant to the developed world, and particularly to Britain. The Summary went on to describe,

> "approaches for effectively applying science, technology and innovation to achieving the Goals. It outlines key areas for policy action, including focusing on generic technologies; improving infrastructure services as a foundation for technology; improving higher education in science and engineering and redefining the role of universities; promoting business activities ion science, technology and innovation; improving the policy environment; and focusing on areas of underfunded research for development."

Two more papers came from this work. 'Going for Growth' arose from a conference organised by the Smith Institute at No. 11 Downing Street. In the Introduction to papers by Juma, he said that, 'the dawn of the new millennium has offered humanity the opportunity to reflect on major global issues. The adoption of the UN Millennium Declaration in 2000 marked the beginning of a re-examination of international development cooperation, the most elaborate outcome of which is reflected in 'Our Common Interest' — the report of the Commission for Africa chaired by UK Prime Minister Blair. One of the central messages of the report is its emphasis on building Africa's capacity to solve its own problems. This focus is reflected in the stress placed on economic growth as a critical basis for addressing poverty. This collection of essays seeks to elaborate on this theme by underscoring the role of science, technology and innovation in development in general, and in international cooperation in particular. The different chapters signal the growing interest in making the transition from short-term relief-based activities to long-term development based on building competence at all levels of science.'

Chapter 5, 'Infrastructure, Innovation and Development' (Ridley T *et al.*, 2005), opened

> "The absence of adequate infrastructure services is one of the main problems that hinder efforts to develop Africa. Technology and innovation are the engines of economic growth. With the globalisation of trade and investment, technological capabilities are a source of competitive advantage. While infrastructure development and technological development are two of the most important areas of development policy, practitioners and

academics alike tend to consider them as separate issues. The focus of infrastructure development in recent years has shifted from merely construction of physical facilities to appropriate provision of services. Environmental and social factors have become part of infrastructure development and planning. Yet most infrastructure projects are not explicitly linked to technological development efforts. The aim of this chapter is to show that infrastructure development can contribute greatly to technological development. It stresses that adequate infrastructure is a necessary requirement for enhancing the creation and application of science and technology in development. Infrastructure development also serves as a technological learning process, which provides individuals, firms and governments with opportunities to acquire and diffuse new knowledge and skills."

This was followed by a paper of the same title (Ridley T *et al.*, 2006) in the *International Journal of Technology and Globalisation.*

The final contribution to WFEO came in 2010, with the publication of the UNESCO Report on Engineering. Tony Marjoram, who had been part of WFEO for many years and who was the sole engineer in the UNESCO hierarchy invited me to write three pieces in the document (Ridley T, 2010) that included a statement by Barry Grear, President 2007–2009, about WFEO,

"This Report presents an important opportunity. As the first ever international report on engineering, it gives the world's engineering community a chance to present the significant contribution that engineering makes to our world. ... The concerns, ideas and examples of good practice captured in the Report provide valuable information for government policy-makers, engineering organisations, international development organisations, engineering colleagues and the wider public to understand the future of engineering, capacity needs, engineering and technical education and engineering applications."

So ended my international professional life. I had 'thought international', starting with my early burning desire to go to the States in 1955. Happily, I had been given the opportunity to make a small contribution to the profession, and to people in other countries around the world.

## Bibliography

CEC (1996), Engineering to survive — global solutions for sustainable development, In *Proceedings CEC 50th Anniversary Inaugural Conference*, London.

Juma C and Y C Lee (2005), Innovation — applying knowledge in development, Task Force on Science, Technology and Innovation, UN Millennium Project, London, Earthscan.

Lee Y C (2010), *Think Malaysian Act Global*, Academy of Sciences, Malaysia.

NCE (2003), Council puts sustainability top of education curriculum, 2/9 Jan.

Ridley T (2010), in UNESCO, Engineering, Engineering and technology in the third millennium (3.5), International cooperation (4.3.2), Transportation (6.2.4).

Ridley T *et al.* (1997), A vision for transport 2020, Engineering Council.

Ridley T *et al.* (2005), Infrastructure, innovation and development, in *Going for Growth: Science, Technology and Innovation in Africa*, Juma C (ed.), Smith Institute, London.

Ridley T *et al.* (2006), Infrastructure, innovation and development, In *International Journal of Technology and Globalisation*, Vol. 2, No. 3/4, pp. 268–278.

Thom D (1993), The new technical culture, CEC conference, Kingston , Jamaica.

Thom D (2014), Onset of the environmental age, New Zealand. IBSN 978-0-473-30182-8.

## Personal Correspondence

Gimson S to T Ridley (19 March 1996).

# Epilogue

## The Role of Engineers

Engineers and engineering have a great capacity to do good for society, though our output is not always universally welcome. Yet we have been mistaken in taking too much *pride in our products per se* and too little in the *services*, such as water and transport services, that our products deliver.

The central role of engineering in society and the economy is evident neither to the public nor the media. The universe of engineering is much larger than is generally supposed. One or more of the engineering disciplines is involved to a significant degree, for example — from agriculture and food to transport, from leisure and entertainment to surgery. We *make things happen and get things done*. Implementation is essential but we must do the *right things well*.

Civil engineers are responsible for much of the essentials of modern life — the muscles and sinews which hold our society together (bridges, roads, railways, dams, airports, docks and tunnels); the provision and maintenance of its heart and lungs (clean water, natural resources in and waste out); transport for safe and effective movement; energy to make it all work (offshore gas and oil, nuclear, hydro, tidal and wind power). Involvement with railways taught me the futility of engineers living and thinking in 'silos'. Many skills are required to *plan, design, finance, build and operate* an urban railway, including mechanical and electrical as well as civil engineering.

When Bazalgette died in 1891 an obituary paid tribute to his contribution to 'the sanitary improvement and to the stateliness of London'. I have no doubt that engineers did as much to improve the health of Victorian Britain as did the medical profession. This is not to minimise the contribution of doctors — simply to say that prevention is at least as important as cure.

When 2013 was celebrated as the 150th anniversary of London Underground a meeting of one of Imperial College's metro benchmarking clubs (described in Chapter 10) was held in London. When addressing the members I observed that, on returning to London from Hong Kong in 1980, I had left behind the world's first modern Chinese metro. Among the world metros attending in 2013 were Hong Kong *plus* Beijing, Shanghai, Guangzhou and Taipei, with Nanjing to come. By the end of that year 10 Chinese cities had operational metro systems, now many more. Not only did the London pioneers in 1863 set in motion something that *changed the face of London for ever*, they did much more — they provided the catalyst that *changed the face of most major cities in the world*.

## The Breadth of Engineering

I discovered the breadth of engineering in my first year at University, courtesy Hardy Cross and two ICE Presidents. Every project of my career has illustrated the crucial importance of the concept of that breadth.

The case studies in the book have illustrated the many skills necessary for successful engineering projects, and lead to a series of propositions. Every major project needs a clear statement about *what we are trying to achieve*, based on a *strategy* with thought-through *objectives*, that is *affordable* (which allows for subsidies as a result of overt public policy). In *implementation*, it is essential that a project should have sound *decision-making and management processes*, focussed on both *delivery* and *operational success*.

It might seem obvious that these principles are essential, but far too many projects have failed, or have been less successful than they might have been, because they have been ignored. No matter how good the consultants, contractors, financiers and others involved, the central requirement is for an *intelligent client* (or *owner*) and, desirably, a personal *champion* of the project. However, while engineers should follow good planning processes, they will occasionally face *windows of opportunity* that are quite fortuitous, when grasping a nettle is necessary because *the time is ripe*.

*Hard science and engineering* are *necessary*, but *not sufficient*, skills. The chemistry of relationships is important too. Clients and contractors *win together or lose together*. I know of no successful intermediate position.

At last, the profession is beginning to recognise, but not yet universally practice, the need to focus on *outcomes* not just outputs, on project planning from *objectives to project control,* on effective *risk management,* on clear *accountability* and short lines of communication, on *lifetime performance* rather than just the creation of infrastructure. We have not recognised that projects frequently fail because of *insufficient investment* in preparation — particularly adequate preparation of the investment proposal *before* committing to delivery. *Unrealistic* schedules, *unrealistic* post-commitment budget reductions, and ignoring the need to have the *operator* as part of the design process, have frequently led to project failure.

I conclude that the most risky stage of an engineering project is at the 'front-end', i.e. when the objectives are not yet established, and the project plan and the culture of the project team are still being developed. The saddest epitaph of a failed project surely must be, '*if only we had spent more time at the front-end*'.

## New Challenges

The need to create resilient and sustainable low-carbon infrastructure systems from existing networks that are variously aged, congested and of low-technology, is increasingly understood. New technologies are becoming available, witness sophisticated methods of monitoring ground movements that have advanced the contribution of geotechnical engineers in tunnelling. This transformation poses challenges of great complexity and requires the best minds and motivation to *contribute to policy* and effect necessary changes.

Meanwhile, engineering journals are full of discussion of new challenges, *inter alia* to

- *Reduce the cost* of projects.
- Develop projects with *predictable outcomes.*
- *Integrate* infrastructure projects into their environment.
- Create *resilient* infrastructure networks, that can withstand environmental turbulence and threats.
- *Upgrade, enhance and replace* infrastructure systems while maintaining effective operations.
- Create *sustainable construction businesses*, both contractors and consultants.

- *Shorten* the supply chain.
- Use IT and data-sharing to step-change *design and construction processes*, and to *communicate* with customers.
- *Focus* on the long-term planning of projects, enabling *dialogue* with beneficiaries (and losers) and sponsors, and the development of *funding* strategies.

Even this short list indicates that engineering presents greater challenges than in the past. Engineers must be widely creative, in *all areas* of appropriate skills — *ingenerare* is Latin for 'to create'.

Another challenge of a different dimension is presented by the forecast retirement rate of ageing engineers. But every threat is also an opportunity. The good news for students and young engineers is that there will be increasing numbers of good jobs available — particularly for the multi-skilled.

## 21ˢᵗ Century Engineers

What then are 21ˢᵗ century engineers? They must be different from those of the late 20ᵗʰ century, and very different from those of the mid-20ᵗʰ century. I offer below, not a definition, but a 'think-piece' to stimulate debate.

Engineers must understand, and help to develop, the *purpose* of major projects and their benefits and dis-benefits. They need to

- *Participate* in the planning process in the round.
- Be able to *communicate* with decision makers.
- Be *advocates* in public debate.
- *Encourage* sustainable decisions.

Their aim should be to lead or assist in the *creation* of relevant projects that are successful throughout their operational life.

Essential capabilities include the capacity to *get* (*the right*) *things done*, which is much more than technical capacity. It requires the ability to *influence others* throughout the *whole life of a project*

- At the *beginning* (what shall be done and why, and where is finance to come from).
- Through what is normally understood as the engineering stage of design and construction.

- To the *final* stage of an operational facility that meets the requirements of *performance* and meets the *objectives* of the project.

Engineers are normally capable of *delivering the second*, but all too rarely do they understand the *first and final stages*. The more they experience all *three* stages the more successful they will be.

To exercise influence engineers must also understand the agendas of *clients and other stakeholders*, to understand 'where they are coming from', and to see the project from their point of view — hence, an appreciation of the *interpersonal*.

None of the above is to decry the traditional specialist, whether designer or constructor, or to fail to recognise that academics are driven by the *merciless pressure* of survival in their special areas of research. But the words of my Australian electrical engineer niece are appropriate. 'The problem with the engineering profession is that we have not recognised that many of the skills required for success have *nothing to do with either maths or physics*'. Ouch, but it is true. Of course maths and science-based knowledge is very important, but more is required of today's engineer for the delivery of successful infrastructure services.

It sometimes seems as though our maths and physics, rather than *enabling us to follow our vocation of providing services for our customers in society*, in fact *act as a mental straight-jacket which forces us to concentrate only on the 'things' we produce*.

## Challenges for Educators and Employers

All of this requires serious debate. The difficulty is not the above capabilities themselves, but how and when the necessary knowledge can be imparted in existing crowded undergraduate and/or postgraduate education programmes. It will be for educators (working with employers) to decide whether and when the knowledge is best imparted through courses, options, reading lists, student essays, student projects in teams, visiting practitioners, or some combination of them. The professions, in the case of ICE through their role as members of the Joint Board of Moderators, should also be involved.

It is unlikely that students could learn, within a four-year degree, everything that they will need to know in order to perform adequately, let alone successfully. While they develop their technological skills, they are sometimes weak in *management and communication skills.* It is not enough to pick these up in the *haphazard* fashion of my and subsequent generations.

Our ambition should be that all engineering students, particularly the high-flyers, at the time of their graduation should, *at the very minimum, be aware* of the many skills that they might need, and are likely to have to use, during their careers.

While much progress has been made since the 1990s, when I was involved in the education and training of young engineers, educators and employers of tomorrow's engineers have not yet sufficiently inculcated and provided the experiences that enable young engineers to fulfil their potential; this at a time when the need for their contribution has probably never been greater. But there is still a gap, the lack of understanding of the *totality* of the engineer's contribution to the major projects that we manage.

If this is accepted then it follows that engineers need an *holistic understanding* of the world they work in and are required to influence. This sets a profound challenge

- For *educators* to contribute to the understanding by young engineers of the breadth of engineering, and
- For *employers* to provide necessary breadth of knowledge, and/or experience, in *planning* (in its totality), *design, implementation and operations.*

Students and new graduates themselves should, by moving within or between employers,

- Seek opportunities for a breadth of experience (answering *why* as well as *what* questions).
- Encourage advice from mentors (never be afraid to ask).
- Seek and learn lessons from elsewhere (every project provides *new insights*).
- Increasingly seek *world-wide* experience.

Thus, it is essential for educators (as providers) and employers (as customers) to work more closely together and with the Professions, to develop an increasing number of young engineers (themselves customers) and thus to meet the ever growing needs of society in the 21$^{st}$ century.

I am very clear that the above is true for engineers in the construction industry, although I learned much about mechanical and electrical engineering through involvement with railway projects, engineers from those disciplines might want to put their own slant on what I have said. However I would not change a word of what I have written about the breadth of engineering.

Finally, although transport, in its widest sense, only 'came in from the cold' in ICE at the time when I became a VP, its particular characteristics enhance understanding about the breadth of engineering. Lessons learned from transport have much to teach us about civil and other engineering disciplines.

*Transport is politics*, but not only transport projects require a united political stance for success. What transport has brought to Civil Engineering is a profound understanding that the *demand* side of the equation is as important as the *supply* side. Indeed, experience suggests that *inaccurate revenue forecasts* can be much more catastrophic than *inaccurate cost forecasts*. Furthermore, Civil Engineering requires a detailed understanding of various types of materials. But Transport demands an understanding of the most difficult material of all — *people*.

My years as a practising engineer and an academic have convinced me that engineering is *a yet broader discipline* than I, or the public and politicians, have recognised — even into the 21st century. This central fact has *direct consequences* for the educators and employers of *tomorrow's engineers*. The profession will be an *even more important, influential and exciting place* to be than it has been in the past. The words of *Hardy Cross, Hindley and Inglis*, would be a good place to start.

# Glossary

| | | |
|---|---|---|
| ACE | — | Association for Consultancy and Engineering |
| AMEME | — | Association of Mining, Electrical and Mechanical Engineers |
| AMICE | — | Associate Member of the Institution of Civil Engineers |
| APM | — | Association for Project Management |
| ASCE | — | American Society of Civil Engineers |
| ASLEF | — | Associated Society of Locomotive Engineers and Firemen |
| ATO | — | Automatic Train Operation |
| BA | — | Booz Allen |
| BEPIC | — | Built Environment Professions in the Commonwealth |
| BR | — | British Rail |
| CEC | — | Commonwealth Engineers Council |
| CEGB | — | Central Electricity Generating Board |
| CIHT | — | Chartered Institute of Highways and Transportation (previously IHE and IHT) |
| CILT | — | Chartered Institute of Logistics and Transport (previously CIT) |
| CHOGM | — | Commonwealth Heads of Government Meeting |
| CSD | — | Commission for Sustainable Development (of the United Nations) |
| CTS | — | (HK) Comprehensive Transport Study |

| | | |
|---|---|---|
| DfT | — | Department for Transport (and variously DoT and DTp) |
| DG | — | Director General |
| DIPTRANS | — | (HK) Development of an Integrated Public Transport System |
| DLR | — | (London) Docklands Light Railway |
| DTI | — | Department of Trade and Industry |
| E&M | — | Electrical and Mechanical |
| E&SB | — | (ICE) Engineering and Sustainability Board |
| EC | — | Engineering Council |
| ERM | — | Enterprise Risk Management |
| ESU | — | English Speaking Union |
| ET | — | Eurotunnel |
| Exco | — | (HK) Executive Council |
| FEANI | — | European Federation of National Engineering Associations |
| FF&A | — | Freeman Fox and Associates |
| FF&P | — | Freeman Fox and Partners (sometimes FF) |
| FFWS&A | — | Freeman Fox, Wilbur Smith and Associates |
| FMC | — | (HK) Finance Management Committee |
| FS | — | (HK) Financial Secretary |
| GA | — | General Assembly |
| GIS | — | Geographical Information System |
| GLC | — | Greater London Council |
| GLDP | — | Greater London Development Plan |
| GoI | — | Government of Indonesia |
| HFA | — | Halcrow Fox and Associates |
| (HK) MTRC | — | Hong Kong Mass Transit Railway Corporation |
| (HK) MTRPA | — | Hong Kong Mass Transit Railway Provisional Authority |
| (HK) MTS | — | Mass Transit Study |
| HND | — | Higher National Diploma |
| HoD | — | Head of Department |

| | | |
|---|---|---|
| IAESTE | — | International Association for the Exchange of Students for Technical Experience |
| ICE | — | Institution of Civil Engineers |
| ICL | — | Imperial College London |
| ICSU | — | International Council of Scientific Unions |
| IEE | — | Institution of Electrical Engineers (now IET — Institution of Engineering and Technology) |
| IMechE | — | Institution of Mechanical Engineers |
| IOC | — | International Olympic Committee |
| IPG | — | (ICE) Infrastructure Policy Group |
| ITTE | — | Institute for Transportation and Traffic Engineering (later ITTS) |
| K&D | — | Kennedy and Donkin |
| KCR | — | Kowloon and Canton Railway |
| LDDC | — | London Docklands Development Corporation |
| Legco | — | (HK) Legislative Council |
| LRPC | — | London Regional Passengers Committee |
| LRTA | — | Light Rail Transit Association |
| LT | — | London Transport (and variously LRT, LTE, TfL) |
| LTRS | — | (HK) Long-Term Road Study |
| LTS | — | London Transportation Study |
| LU | — | London Underground (and variously LUL) |
| MD | — | Managing Director |
| MDGs | — | (UN) Millennium Development Goals |
| MIS | — | (HKMTR) Modified Initial System |
| MPA | — | Major Projects Association |
| MRTA | — | (Bangkok) Metropolitan Rapid Transit Authority |
| NAO | — | National Audit Office |
| NCE | — | *New Civil Engineer* |
| NESDB | — | (Thailand) National Economic and Social Development Board |
| NPPC | — | Nuclear Power Plant Company |

| | | |
|---|---|---|
| NTU | — | Nanyang Technological University (Singapore) |
| NUR | — | National Union of Railways (now RMT) |
| NUS | — | National University of Singapore |
| ODA | — | Olympic Delivery Authority |
| OGGS | — | One Great George Street (ICE Headquarters) |
| OPO | — | One Person Operation |
| OR | — | Operations (or Operational) Research |
| PLBs | — | (HK) Public Light Buses |
| PM | — | Project Manager (of DLR) |
| PWD | — | (HK) Public Works Department |
| RAC | — | Royal Automobile Club |
| RAMP | — | Risk Analysis and Management for Projects |
| RIA | — | Railway Industry Association |
| RRS | — | (HK) Revenue Review Study |
| RSA | — | Railway Studies Association |
| RSA | — | Royal Society of Arts |
| RTSC | — | Railway Technology Strategy Centre (at ICL) (now Railway and Transport Strategy Centre) |
| SAICE | — | South African Institution of Civil Engineers |
| SG | — | Steering Group |
| SCMP | — | South China Morning Post |
| SMRT | — | Singapore Mass Rapid Transit |
| SVP | — | Senior Vice President |
| T&WPTA | — | Tyne and Wear Passenger Transport Authority |
| T&WPTE | — | Tyne and Wear Passenger Transport Executive |
| TE&C | — | Traffic Engineering and Control |
| TfL | — | Transport for London |
| TGWU | — | Transport and General Workers Union |
| TML | — | Transmanche Link |
| TNPG | — | The Nuclear Power Group |
| TOR | — | Terms of Reference |

| | | |
|---|---|---|
| TRL | — | Transport Research Laboratory (previously RRL and TRRL) |
| TUBE | — | London Underground |
| UCL | — | University College London |
| ULCTS | — | (University of London) Centre for Transport Studies |
| UITP | — | International Union of Public Transport |
| UTS | — | Underground Ticketing System |
| WFEO | — | World Federation of Engineering Organizations (GA — General Assembly and Exec — Executive Council) |

# Index

CPSIA information can be obtained
at www.ICGtesting.com
Printed in the USA
LVOW05*1925040318
568598LV00004B/19/P